I0065323

Photonics for Radar Networks and Electronic Warfare Systems

Related titles on radar

Advances in Bistatic Radar Willis & Griffiths
Airborne Early Warning System Concepts, 3rd Edition Long
Bistatic Radar, 2nd Edition Willis
Design of Multi-Frequency CW Radars Jankiraman
Digital Techniques for Wideband Receivers, 2nd Edition Tsui
Electronic Warfare Pocket Guide Adamy
Foliage Penetration Radar: Detection and characterisation of objects under trees Davis
Fundamentals of Ground Radar for ATC Engineers and Technicians Bouwman
Fundamentals of Systems Engineering and Defense Systems Applications Jeffrey
Introduction to Electronic Warfare Modeling and Simulation Adamy
Introduction to Electronic Defense Systems Neri
Introduction to Sensors for Ranging and Imaging Brooker
Microwave Passive Direction Finding Lipsky
Microwave Receivers with Electronic Warfare Applications Tsui
Phased-Array Radar Design: Application of radar fundamentals Jeffrey
Pocket Radar Guide: Key facts, equations, and data Curry
Principles of Modern Radar, Volume 1: Basic principles Richards, Scheer & Holm
Principles of Modern Radar, Volume 2: Advanced techniques Melvin & Scheer
Principles of Modern Radar, Volume 3: Applications Scheer & Melvin
Principles of Waveform Diversity and Design Wicks *et al.*
Pulse Doppler Radar Alabaster
Radar Cross Section Measurements Knott
Radar Cross Section, 2nd Edition Knott *et al.*
Radar Design Principles: Signal processing and the environment, 2nd Edition Nathanson *et al.*
Radar Detection DiFranco & Ruby
Radar Essentials: A concise handbook for radar design and performance Curry
Radar Foundations for Imaging and Advanced Concepts Sullivan
Radar Principles for the Non-Specialist, 3rd Edition Toomay & Hannan
Test and Evaluation of Aircraft Avionics and Weapons Systems McShea
Understanding Radar Systems Kingsley & Quegan
Understanding Synthetic Aperture Radar Images Oliver & Quegan
Radar and Electronic Warfare Principles for the Non-specialist, 4th Edition Hannen
Inverse Synthetic Aperture Radar Imaging: Principles, algorithms and applications Chen & Marotella
Stimson's Introduction to Airborne Radar, 3rd Edition Baker, Griffiths & Adamy
Test and Evaluation of Avionics and Weapon Systems, 2nd Edition McShea
Angle-of-Arrival Estimation Using Radar Interferometry: Methods and applications Holder
Biologically-Inspired Radar and Sonar: Lessons from Nature Lessons from Nature Balleri, Griffiths & Baker
The Impact of Cognition on Radar Technology Farina, De Maio & Haykin
Novel Radar Techniques and Applications, Volume 1: Real Aperture Array Radar, Imaging Radar, and Passive and Multistatic Radar Klemm, Nickel, Gierull, Lombardo, Griffiths and Koch
Novel Radar Techniques and Applications, Volume 2: Waveform Diversity and Cognitive Radar, and Target Tracking and Data Fusion Klemm, Nickel, Gierull, Lombardo, Griffiths and Koch
Radar and Communication Spectrum Sharing Blunt & Perrins
Systems Engineering for Ethical Autonomous Systems Gillespie
Shadowing Function from Randomly Rough Surfaces: Derivation and applications Bourlier & Li

Photonics for Radar Networks and Electronic Warfare Systems

Edited by
Antonella Bogoni, Paolo Ghelfi and Francesco Laghezza

The Institution of Engineering and Technology

Published by SciTech Publishing, an imprint of The Institution of Engineering and Technology, London, United Kingdom

The Institution of Engineering and Technology is registered as a Charity in England & Wales (no. 211014) and Scotland (no. SC038698).

First published 2019

The Institution of Engineering and Technology
Michael Faraday House
Six Hills Way, Stevenage
Herts, SG1 2AY, United Kingdom

www.theiet.org

British Library Cataloguing in Publication Data
A catalogue record for this product is available from the British Library

ISBN 978-1-78561-376-0 (hardback)
ISBN 978-1-78561-377-7 (PDF)

Typeset in India by MPS Limited
Printed in the UK by CPI Group (UK) Ltd, Croydon

Contents

About the authors

Antonella Bogoni, head of research area of CNIT (National Inter-University Consortium for Telecommunications), is one of the pioneers of the Integrated Research Center for Photonic Networks and Technologies created in Pisa in 2001 as a part of a broad agreement among four partners, CNIT, Scuola Superiore Sant'Anna, Marconi Communications S.p.A. (now Ericsson), and the National Research Council of Italy (CNR), aiming to create in Pisa a world-class Center of Excellence, where currently she is leader of the œdigital & microwave photonics area. Antonella Bogoni dedicated her research activity to photonics technologies for ultra-fast optical communication systems in order to meet the needs of a faster Internet and of a new generation of user services. In 2004 she proposed the fastest optical regenerator at that time, working at 160 billions of bit per second thanks to photonics technologies (10 times faster than the electronic technologies available at that time). Also in 2004 she obtained her first national funding for the realization of a new prototype of pulsed laser for high-capacity optical communication networks. From 2008 she collaborates with the University of Southern California (USC) in Los Angeles where she worked in 2008 and from 2009 to 2010 as winner of a Fulbright scholarship. In the USC laboratories she developed the first photonic processor prototype suitable for logic operations using the light with a speed of 640 billions of operations per second, about one hundred times higher compared to the electronics. In 2009 she was scientific person in charge for CNIT of the project ACEPLAN ACtivE PLAsmoNics and lossless metamaterials funded by European Consortium NanoSci-ERA. In 2009 she obtained an ERC starting grant for developing a photonic-based fully digital radar systems, and in 2012 she got a second ERC grant within the proof-of-concept program in order to convert her research results into a pre-industrial product for airport security. Also in 2012 Antonella Bogoni started collaborating with the Italian Defence Ministry that supports her group activities for the development of a photonic-based TLC radio system of military interest. In 2006 she was co-founder of PhoTrix, a start-up company manufacturing and selling ultra-high-speed optical instrumentation; from 2007 to 2011 she was CEO of PhoTrix. Up to now Antonella Bogoni has got 11 national and international project grants, she is co/author of 45 patents, 8 books, and more than 110 papers on the main scientific international journals. Moreover, she presented more than 200 contributions in the main photonics microwave photonics and radar international conferences. She has been also invited as expert on the main journals and conferences and she collaborates with several of the most prestigious research and industrial groups in photonics and radar fields all around the world, that is, Europe, the USA, Brazil, Japan, Korea, etc. She also participated in the technical committee and advisory board for international conferences and journals respectively. In 2014 she was general co-chair of the International IEEE Conference Photonics in Switching. Her main current interests are in microwave photonics, ultra-fast optical communications, photonic digital processing, and nonlinear optics.

Paolo Ghelfi received the M.S. degree in electronics engineering from the University of Parma, Italy, in 2000. From 2001 he is with the National Laboratory of Photonic Networks of CNIT in Pisa, where he is now Head of Research. In 2008 he has been visiting scientist at KIST in Seoul, South Korea. With the establishment of PhoTrix Srl (2006–2011), a spin-off company of CNIT and Scuola Superiore Sant'Anna, he experienced in the management of a company, dealing in particular with administration and production. His research interests are in the field of optical communications, in particular in all-optical communication systems and digital photonics (pulsed laser sources, OTDM systems and all-optical processing, 3R all-optical regeneration, all-optical networking), and on WDM systems and networks (PMD compensation techniques, WDM-PON networks and access networks, reconfigurable optical networks). Recently he has been exploring the topic of microwave photonics, in particular for applications to radar systems (optical generation of ultra-stable RF carriers over wide frequency ranges, optical generation of wideband amplitude- and phase-modulated RF signals, precise optical sampling of RF signals, processing of optical samples for high-sampling-rate RF receivers, optical beamforming and true-time delay, radio-over-fiber) and to wireless communications systems (optical up- and down-conversion of reconfigurable wideband RF signals). From 2009 he acts as laboratory coordinator for the activities on microwave photonics. Paolo Ghelfi has served as lecturer in several courses of international masters and Ph.D. programs held by Scuola Superiore Sant'Anna, Pisa, Italy. He also serves as technical contact for collaborations between the LNRF and the several companies, as Selex ES, IDS SpA, Elettronica SpA, Consilium Italy, and Caliopa. He has authored or co-authored 27 papers on international journals, more than 85 papers on international conferences, and 14 patents.

Francesco Laghezza is a Post Doc Researcher at the Signal Processing Systems (SPS) Group, Department of Electrical Engineering, Eindhoven University of Technology (TU/e) in The Netherlands. His main research focuses on joint radar communication aspects and challenges for the automotive scenario. Dr. Laghezza has authored more than 70 papers in leading peer-reviewed journals and conferences, and 5 patents. During his activities, he was honored with the Young Engineering prize, during the 2014 International Radar Symposium, and with the Thales prize for his contribution in developing the first hybrid, LIDAR\RADAR integrated architecture. From 2010 to 2016, he was a researcher at CNIT (National Interuniversity Consortium for Telecommunications) in Pisa (Italy), working as technical leader for the radar system developing and prototyping group within the European projects PHODIR, PREPARE, and PETRA. He also worked for the development of a Microwave-Photonic Dual Band Radar (DBR) and a fiber based distributed RADARCOM network, a photonic receiver for Electronic Spectrum

Measurement (ESM) and SIGnal INTelligence (SIGINT) in close collaboration with national and international companies, research bodies, and DoD. His main activities are in the area of research and development of radar systems with more than 7 years' experience in system design, system engineering hands-on capability, including system and structural design, field test, and calibration.

Preface

In the last decades, the use of photonics has been proposed for improving several microwave subsystems by solving very specific problems as, for example, reducing radio frequency (RF) propagation losses, or increasing the bandwidth in RF detection. In the technical literature, photonics was spread over several different applications. The most mature was the signal transportation (usually called radio-over-fiber or microwave photonic link), but numerous examples of microwave photonic filters, photonics-based RF signal generators, or photonics-enabled high-sampling-rate analog-to-digital converters, to name a few different subsystems, had been proposed.

Instead, a new research approach is recently rising, investigating the development of full photonics-based systems by taking advantage of all the special features of photonics in a single apparatus. Although still at an embryonic stage, few photonics-based systems have been presented, implementing in particular radars, radar networks, or electronic warfare (EW) systems.

Moreover, the advances in photonic integration are making several ideas closer to a real implementation and deployment, thanks to strongly reduced encumbrance and strongly increased stability and reliability.

Working ourselves in this scientific environment, we felt the need of realizing a book to report the potentials of microwave photonics in a more homogeneous, complete, and easy-to-read way, with the final aim of fostering an industrial exploitation of the demonstrated proof of concepts.

Therefore, the main goal of this book is the delivery of detailed and clear interdisciplinary information, at both research and industrial levels, useful for the involved scientific communities: the radar and EW system community on one side, and the photonics community on the other side, which are usually accustomed to different languages and approaches.

At the same time, the aim of the book is also to bring these two different worlds closer, promoting their cross-fertilization and boosting both the development of new photonic technologies for innovative applications, and the improvement of radar and EW systems' performance.

Therefore, this book can be of help for, on one hand, designers and researchers dealing with radar and EW systems who want to understand, explore, and make use of microwave photonics concepts, and on the other hand, researchers and engineers in photonics who want to have a complete picture of the systems where their components or chips might be used.

The book describes the main hardware functionalities provided by photonics in radar and EW systems: RF transport in optical fiber, photonics-based RF signal generation/up-conversion and analog-to-digital conversion/down-conversion, optical beamforming, and optical RF filtering. The book also describes the new radar and EW system architectures enabled by photonics and their impact on the digital signal processing. In addition, the capabilities of photonics integration are addressed, highlighting its potential in reducing the size, weight, power consumption, and cost of the whole radar or EW systems, and reporting on the new applications allowed by the on-chip system implementations.

In more detail, the book is organized as follows.

First, the structure and figures of merit of radar systems (Chapter 1) and EW systems (Chapter 2) are summarized, highlighting in particular their current issues. Then, Chapter 3 introduces the building blocks of microwave photonics, underlining the potentials of each described photonics-implemented function, while Chapter 4 reports on the recent examples of photonics-based radar systems. The general concepts and figures of merit of radar networks (or netted radars) are described in Chapter 5, reporting also on the examples of radar networks implemented with standard RF technologies and highlighting their limitations. On the other hand, Chapter 6 describes the potentials of very recent implementations of photonics-based coherent radar networks. Instead, the most relevant examples of photonics in EW systems are reported in Chapter 7. Finally, Chapter 8 reports an analysis of the past and future of photonics in radar and EW systems from an industrial viewpoint.

To author the chapters above, we have invited a group of world-renowned experts in each of the involved fields: radars, radar networks, EW systems, and microwave photonics, who have been or are currently collaborating on these new applications of photonics at system level. This approach ensures the chapters share a common line, and the entire book holds a unitary viewpoint. Moreover, the invited authors show a balanced background of academic and industrial expertise, which gives the book a critical vision merging the scientific potentials and exciting novelties described by an academic approach, with the solid analysis based on technical feasibility and cost/performance balancing required by the actual market needs.

Antonella Bogoni
Paolo Ghelfi
Francesco Laghezza

Chapter 1
Issues on current radar systems

Fabrizio Berizzi[1], Amerigo Capria[2], Elisa Giusti[2] and Anna Lisa Saverino[2]

Acronyms

ACC	Autonomous cruise control
ACF	Automatic frequency control
ADC	Analog to digital converter
ADT	Automatic detection and tracking
COHO	Coherent oscillator
CPA	Closest point of approach
CFAR	Constant false alarm rate
CW	Continuous wave
DDS	Direct digital synthesizer
E.M.	Electromagnetic wave
ENOB	Equivalent number of bits
FB	Feature based
HF	High frequency
HRRP	High-resolution range profiles
I.F.	Intermediate frequency
ISAR	Inverse synthetic aperture radar
LO	Local oscillator
MF	Matched filter
MM	Model matching
MS	Match score
NCTR	Non-cooperative target recognition
OTH	Over the horizon
PLL	Phased locked loop

[1]Department of Information Engineering, University of Pisa, Pisa, Italy
[2]Radar and Surveillance Systems (RaSS) National Laboratory, National Inter-University Consortium for Telecommunications (CNIT), Pisa, Italy

P.R.I.	Pulse repetition interval
RADAR	Radio detection and ranging
RCS	Radar cross section
RF	Radio frequency
SAR	Synthetic aperture radar
SFDR	Spurious free dynamic range
SLAR	Side-looking airborne radar
STALO	Stable local oscillator
STT	Single target tracker
TWS	Track and while scan
UHF	Ultra high frequency
VCO	Voltage-controlled oscillator
VHF	Very high frequency
WSR	Weather surveillance radar

1.1 Chapter organization and key points

Radio detection and ranging (RADAR) detects and locates targets within a limited volume in space by transmitting electromagnetic (E.M.) energy and processing the reflected echoes. Other target characteristics such as velocity shape and size can be extracted as well. Despite the concept of detecting targets of interest using radio waves gaining mainly the interest of the military community, nowadays radars are also used in civil applications like short-term weather forecasting, geological observations, autonomous cruise control, and so on. The diversity of applications calls for numerous radar types, each of which can be classified according to several aspects like specific radar features (e.g., the frequency band, antenna type, transmitted waveform), radar platform (e.g., ground based, airborne, spaceborne), radar mission and functionality (e.g., tracking, early warning, weather), and radar architecture (e.g., monostatic, multistatic, multifunction). As technology and applications advance, radar systems are demanded to face new challenges (e.g., automatic target recognition, radar imaging techniques, capability to work in a network, phase and time synchronization, long-range detection and tracking, etc.). As a consequence, the radar requirements and expectations are becoming demanding bringing to the development of more and more sophisticated radar systems.

This chapter presents an overview of the radar system and its main applications that have pushed on the radar toward new paradigms and system concepts.

In particular, Section 1.2 introduces the basic theory and concepts behind a classical radar system. Specifically, the basic radar nomenclature and architecture are summarized in brief.

Section 1.3 presents a summary of the main radar functionalities and applications.

Sections 1.4 and 1.5 deal with the new challenges that modern radar systems are demanded to face with, and especially a high degree of adaptability to the environment and the capacity to cooperate in a network. Such new features are meant to enhance the radar performance in terms of detection, tracking, and automatic target recognition (ATR) in an dynamic environment and to make radar system able to handle several tasks by managing its resources autonomously and intelligently.

Finally, Section 1.7 analyzes the main issues concerned with the current radar systems to face these new challenges.

1.2 Introduction to radar systems

RADAR indicates a system that uses E.M. waves to determine the presence (detection) and location (targets RADAR distance measurement) of both stationary and nonstationary targets.

The physical principle behind radar functioning relies on the scattering phenomenology. The scattering is the property of a target to backscatter toward the radar a portion of the transmitted radar energy.

From an operational point of view, a radar transmits a certain amount of energy via a radio frequency (RF) signal, with proper characteristics, toward a region of interest, which is usually referred to as the surveillance area. The transmitted signal hits a target in the surveillance area and induces on it the surface currents. The surface currents make the targets a radiating element, which in turn radiates energy in the surrounding area. The way the energy is radiated from the target depends on the surface currents on it, which in turn depend on the E.M. signal transmitted by the radar and the target physical characteristics (basically its shape, size, and the material of which it is made).

A part of the energy backscattered by the target (radar returns or echoes) is reflected back to the radar, which is able to sense this signal and measure its distance by performing a sort of "comparison" between the transmitted signal and the received one. In fact as it can be demonstrated hereinafter, the signal backscattered by the target and received by the radar contains information about the target distance, or range in radar terminology, and target motions with respect to the radar whether the target motions causes a range migration within the time it is observed by the radar.

A typical transmitted signal is a single-frequency pulse defined as follows:

$$s_{\mathrm{T}}(t) = A\mathrm{rect}\left(\frac{t - T_i/2}{T_i}\right)\cos(2\pi f_0 t) \tag{1.1}$$

where A is an amplitude coefficient, $\mathrm{rect}(x) = \begin{cases} 1 & |x| < 1/2 \\ 0 & \text{otherwise} \end{cases}$, T_i is the pulse temporal length, and f_0 is the carrier frequency.

Let us assume a point-like scatterer at a distance R by the radar and moving with a velocity v. Then the signal backscattered by the target and sensed by the radar can be written as:

$$s_{\mathrm{R}}(t) = K\mathrm{rect}\left(\frac{t - T_i/2 - \tau(t)}{T_i}\right)\cos(2\pi f_0(t - \tau(t))) \tag{1.2}$$

where $|K| < |A|$, $\tau(t) = \dfrac{2R(t)}{c}$ is the delay time, and c is the light speed in a vacuum.

The received signal is an attenuated and time-shifted replica of the transmitted signal. As can be noted, the received signal carries information about the target distance over time. It is quite evident that by recovering the information on the range history of the target over time one can estimate the target motion parameters, including target radial velocity and acceleration.

Radar systems can be sketchily categorized into two classes, which differ for the geometry of acquisition:

- Monostatic radars in which the transmitter and the receiver are co-located. In this case,

$$\tau(t) = \frac{2R(t)}{c} \tag{1.3}$$

- Bistatic radars in which the transmitter and the receiver are spatially separated. In this case,

$$\tau(t) = \frac{(R_{\mathrm{T}}(t) + R_{\mathrm{R}}(t))}{c} \tag{1.4}$$

where R_{T} and R_{R} are the transmitted target and receiver target distances.

Radar systems may be also distinguished on the basis of the transmitted waveform among:

- Pulse radars that transmit a pulse as in (1.1), every T_{R} seconds. T_{R} is also referred as the pulse repetition interval (PRI). Pulse radars transmit and receive a signal in different and consecutive temporal slots.
- Continuous waveform (CW) radars that transmit a continuous signal instead of pulses of limited time duration. An example of single-frequency CW waveform is as follows:

$$s_{\mathrm{T}}(t) = A\cos(2\pi f_0 t)\mathrm{rect}\left(\frac{t - T_{\mathrm{obs}}/2}{T_{\mathrm{obs}}}\right) \tag{1.5}$$

where T_{obs} is the observation time. A monostatic CW radar is able to contemporarily transmit and receive a signal.

1.2.1 Radar design

In this section, the basic radar terminology and relations are reviewed in brief. When a radar apparatus must be dimensioned, we should consider the following aspects:

- *Spatial resolution*. It is the minimum distance (along a predefined direction) between two point-like scatterers with equal reflectivity, such that they can be distinguished as two separate objects by the radar system. The range resolution, in particular, is defined as follows:

$$\delta_{\mathrm{r}} = \frac{c}{2B_i} \tag{1.6}$$

where c is the light speed in a vacuum and B_i is the instantaneous bandwidth of the transmitted signal. For single-frequency pulsed radar ((1.1)), B_i is linked to the pulse duration by means of the uncertainty relationship of a Fourier transform pair, therefore, $B_i \propto \frac{1}{T_i}$.

- *Radar coverage*. It is the maximum range at which the radar can detect with a certain rate of detection (probability of detection) a target with a certain reflectivity power (named radar cross section).
- *Maximum nonambiguous range*. In a pulsed radar, this is the maximum range of a target such that the leading edge of the received backscatter echo from that target is received before the transmission of the next pulse.

  The maximum nonambiguous range is then linked to the PRI through the following equation:

$$\Delta_{\mathrm{r}} = \frac{c}{2} T_{\mathrm{R}} \tag{1.7}$$

- *Bind range*. In a pulsed radar, this is the minimum detectable distance of a target. It is limited by the duration of a pulse. In a pulse monostatic radar, in fact a radar cannot receive any echo during the pulse transmission.

  The blind velocity is linked to the pulse duration T_i and is defined as follows:

$$\Delta_{\mathrm{B}} = \frac{c}{2} T_i \tag{1.8}$$

- *Power budget*. In a radar, the power budget establishes the relation between the received and the transmitted power by accounting of all of the gains and losses from the transmitter, through the medium (free space, cable, waveguide, etc.) to the receiver.

The power budget is modeled via the so-called radar equation, which is expressed as in (1.9), where P_{R} is the received echo power, P_{T} is the transmitted power, G_{T} and G_{R} are the antenna gains at the transmitter and receiver, respectively, σ is the target reflectivity coefficient (usually measured in m^2), λ is the transmitted wavelength, R_{T} and R_{R} are the target transmitter and the target receiver distances, respectively, and L are the system (transmitter, receiver, propagation channel) losses.

$$P_{\mathrm{R}} = \frac{P_{\mathrm{T}} G_{\mathrm{T}} G_{\mathrm{R}} \sigma \lambda^2}{(4\pi)^3 (R_{\mathrm{T}} + R_{\mathrm{R}})^2 L} \tag{1.9}$$

Usually, the received echo signal is affected by additive noise, which may affect the radar detection performance.

The radar detection performance depends on the signal-to-noise ratio (SNR), which is in turn linked to the received echo energy E_{s}, and the noise mean power . The SNR is defined as follows:

$$\mathrm{SNR} = \frac{E_{\mathrm{s}}}{P_{\mathrm{n}}} \tag{1.10}$$

In case of a pulse radar, it becomes:

$$\mathrm{SNR} = \frac{P_{\mathrm{R}} \cdot T_i \cdot N}{N_0/2} = \frac{P_{\mathrm{T}} T_i N G_{\mathrm{T}} G_{\mathrm{R}} \sigma \lambda^2}{(4\pi)^3 (R_{\mathrm{T}} + R_{\mathrm{R}})^2 L (N_0/2)} \tag{1.11}$$

where $T_i \cdot N = T_{\mathrm{int}}$ is the observation time.

The detection of a slowly fluctuating point target in presence of additive noise can be formulated as a binary hypothesis problem:

$$\begin{aligned} H_0 &: r(t) = n(t) \\ H_1 &: r(t) = x(t) + n(t) \end{aligned} \tag{1.12}$$

where $n(t)$ is the Gaussian noise signal, $x(t)$ is the target echo signal, $r(t)$ is the received signal, and H_0 and H_1 are, respectively, the null hypothesis and the alternative hypothesis.

To decide whether a target is present or not, a threshold operation is performed on a *decision statistic* that is derived from a transformation of the received signal $r(t)$ as follows:

$$M(r(t)) \gtrless \lambda \tag{1.13}$$

where $M(\cdot)$ is the transformation applied to the received signal that maps the received signal into the *decision statistic*. The *decision statistic* is then compared with a proper threshold, λ.

A commonly used *decision statistic* in radar detection is the absolute (or the absolute squared) value of the *matched filter* (MF) output.

The MF is a low-pass filter that ensures the maximum SNR at its output around the target delay time. The MF impulse response is the reverse conjugate of the transmitted signal, namely:

$$h(t) = s^*(-t) \tag{1.14}$$

Since the received signal is a delayed and attenuated replica of the transmitted signal, the output of the MF is proportional to the autocorrelation function of the transmitted signal (1.15):

$$C_{S_T}(t) = s(t) \otimes s^*(-t) \tag{1.15}$$

The presence of the additive noise determines the decision uncertainties. The radar detection performance is measured by means of both the probability of detection (P_D) and the probability of false alarm (P_{FA}). Both are conditioned probabilities. The probability of detection is the probability to correctly detect a target (e.g., the probability to decide for H_1 under the hypothesis H_1) and is a function of both the SNR and λ as shown in (1.16):

$$P_D = Pr\{H_1|H_1\} = f(\text{SNR}, \lambda) \tag{1.16}$$

The probability of false alarm is instead the probability to detect a target when it is not present in the observed area (e.g., the probability to decide for H_1 under the hypothesis H_0) and is a function of λ, as shown in (1.17):

$$P_{FA} = Pr\{H_0|H_1\} = f(\lambda) \tag{1.17}$$

1.2.2 Radar system architecture

Figure 1.1 illustrates the block diagram of a radar. The radar signal is generated by the transmitter (which is composed of two main blocks, namely the waveform generation and the amplifier) and radiated by the antenna. The duplexer permits a single antenna to be used for both transmission and reception (in case of a pulse waveform). The echo backscattered by the target comes back to the radar, is collected by the antenna, and detected by the receiver. The detection of the echo energy reveals the presence of a target, and the comparison of the received signal with the transmitted one allows target information (target location, size, shape, and velocity with respect to the radar) to be estimated. The results of the signal and data processing are then portrayed on a display [1].

The transmitter should generate the required RF average and peak power and must guarantee a suitable RF stability to meet the signal processing requirements. The transmitters can be sketchily categorized into:

- *Power oscillator transmitter (POT)*. In this system, a power oscillator, usually a magnetron, produces the RF pulses. The power oscillator is keyed by a high-power dc pulse energy generator. Automatic frequency control is often used to keep the receiver tuned to the frequency of the transmitter since the magnetron frequency slowly drifts due to temperature changes.

  Moreover, since the magnetron is an oscillator, the starting phase of each pulse is random from pulse to pulse. This drawback should be properly addressed in the receiver whether the magnetron is used in a coherent radar system. In this case, the phase of the magnetron sets the phase of a coherent oscillator (COHO) at the receiver. In this way, the received signal is coherent from pulse to pulse.

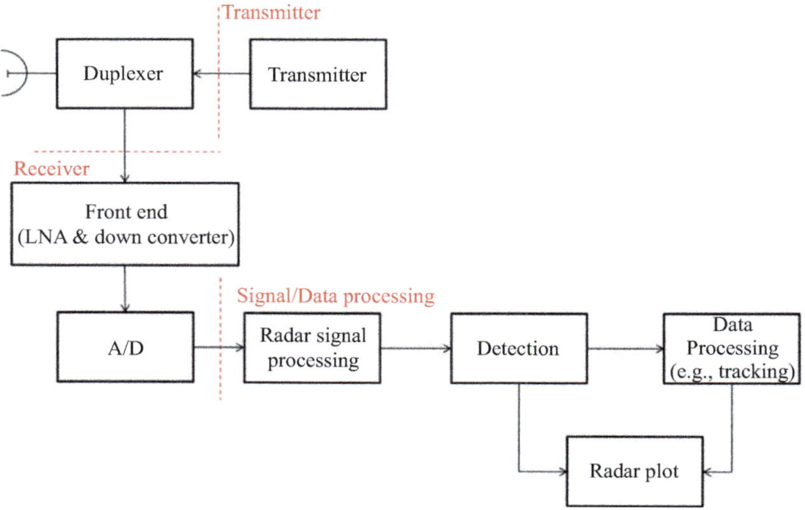

Figure 1.1 Radar system diagram

- *Power amplifier transmitter (PAT)*. In this case, a waveform generator is used to generate the transmitting signal on an intermediate frequency (IF). It permits generating predefined waveforms by driving the amplitudes and phase shifts of carried microwave signals. It is then taken to the necessary power by using an amplifier (amplitron, Klystron, or solid-state amplifier).

Table 1.1 summarizes the main characteristics of these two transmitters.

The main functions of the receiver are (1) to sense the weak echoes through antenna system, (2) to amplify them sufficiently, (3) to detect the pulse envelope, and (4) to feed them to the signal/data processing unit.

All radar receivers employ a super heterodyne architecture, which corresponds to the "front-end" block, in Figure 1.1.

The receiver signal is first amplified via a low noise amplifier, the signal is then shifted to an IF, by mixing with a local oscillator (LO) frequency. More than one conversion stage may be necessary to reach the final IF without facing with serious image or spurious frequency problems. In particular, the LO influences the coherence degree of a radar system.

Incoherent radars do not necessitate to phase coherence. They only need time synchronization, which may be guaranteed via GPS devices and frequency synchronization. If a certain degree of frequency stability is required, the automatic frequency control network is used between the transmitter and the receiver, which corrects the frequency drift of the LO.

Figure 1.2 depicts the common diagram used to schematically represent a coherent radar.

The COHO generates a reference monochromatic signal. The COHO is designed so as to guarantee a suitable frequency stability within the entire processing time, namely $T_{int} = NT_R$. The signal generated by the COHO is up-converted or down-converted to the transmitted frequency or the IF in the transmitter and in the receiver, respectively, through the stable local (STALO) oscillator.

Table 1.1 Technical characterization of typical transmitters [2]

	Technology	Maximum frequency (GHz)	Peak/average power	Gain (typical)	Bandwidth (% of the central frequency)
POT	Magnetron	95	1 MW/500 W@10 GHz	–	10
	Impact diode	140	30 W/10 W@10 GHz	–	5
	Extended interaction oscillator	220	1 kW/10 W@95 GHz	–	Up to 4
PAT	Extended interaction Klystron	280	1 kW/10 W@95 GHz	40–50 dB	Up to 1
	Klystron	35	50 kW/5 kW	30–60 dB	Up to 10
	Solid state silicon (BJT)	5	300 W/30 W	5–10 dB	Up to 15

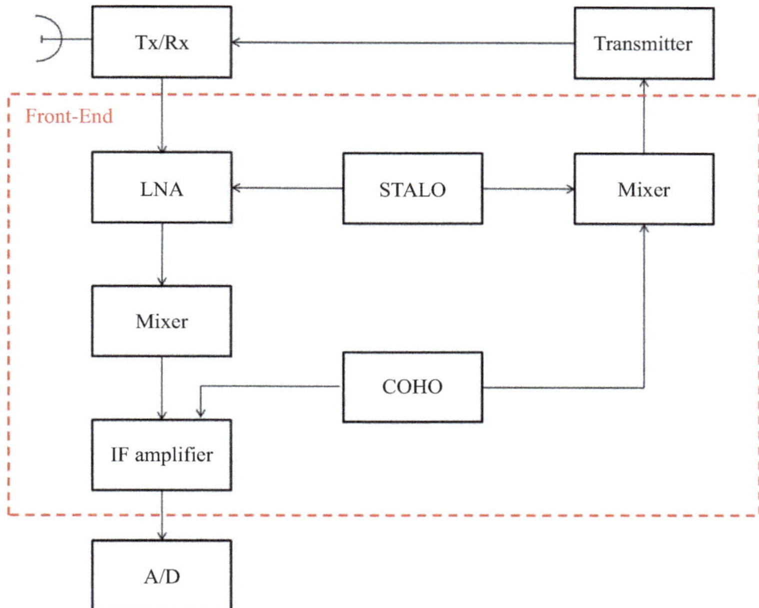

Figure 1.2 Coherent radar front-end

The STALO is required to have a lower degree of frequency stability with respect to the COHO. In fact, the STALO must guarantee a certain degree of frequency stability in a shorter time interval, namely T_R.

The last stage of the front-end is the "IF amplifier," which amplifies the signal at IF.

1.3 Radar functionalities and applications

Radars may be distinguished in terms of their main characteristics, namely the architecture configuration, transmitted waveform, platform, target response, signal processing, frequency band, functionalities, and applications.

The main radar features and the way a radar is classified according to such features are briefly recalled hereinafter:

1. The architecture configuration defines the way transmitters and receivers are located in the space. Examples of radar architectures are monostatic, bistatic, and multistatic radar. In a monostatic radar, the transmitter and the receiver are co-located. Conversely in a bistatic radar system, the transmitter and the receiver are not co-located and a proper time and/or phase synchronization network is necessary to process the received signal.
2. The transmitted waveform represents the way a signal is transmitted through the propagation medium. Based on the waveform, radar may be distinguished among pulsed radar, Doppler radar, CW radar, modulated pulsed radar, and frequency modulated CW radar (refer Figure 1.3).

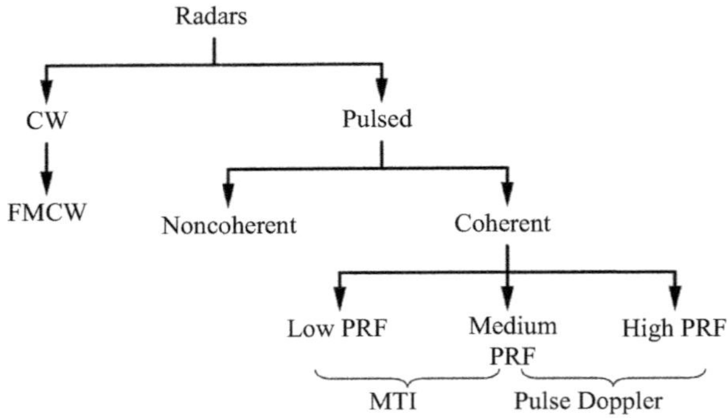

Figure 1.3 Classification by waveform [3]. CW, continuous wave; FMCW, frequency modulated continuous wave; PRF, pulse repetition frequency; MTI, moving target indicator

3. Platform is the support on which the radar system is installed on. Typically used platforms are ground stations, airplanes, and satellites. The radar platform may be stationary (which is typically the case of ground stations) or moving.
4. *Target response*. Based on the target response, radars may be classified between primary radars and secondary radars. Primary radars detect and locate a non-cooperative target based on the reflected target echoes. Secondary radars are instead able to recover additional information about the target besides its location and velocity, such as its identity, altitude, and so on. To do that secondary radars rely on cooperative targets that are equipped with a radar transponder. The radar transponder replies to each interrogation in the signal transmitted to the radar by transmitting a response containing encoded data.
5. *Signal processing* is the way the received signal is processed to extract information on the detected target. The signal processing is constrained by the radar architecture and specifically by the coherence level at the receiver and the antenna system. Based on the signal processing algorithm, a radar system may be defined as coherent, noncoherent, and phased array.
6. *Operating frequency:* High frequency (HF), very high frequency (VHF), ultra high frequency, L, S, C, X, Ku, and Ka bands. Specifically, each operative bandwidth is assigned to accomplish determined radar operative modes. The relation between the band designation and the radar usage is shown in Figure 1.4.

Another way to classify radar systems is based on its functions and applications. Such kind of classification is detailed in the following two subsections.

1.3.1 Definition of radar functionalities

The well-known primary radar functions are detection, search, and tracking. Based on these primary functions, other radar functionalities can be defined, which may enable the system to accomplish a variety of missions and applications.

Band designation	Frequency range	Usage
HF	3–30 MHz	OTH surveillance
VHF	30–300 MHz	Very-long-range surveillance
UHF	300–1,000 MHz	Very-long-range surveillance
L	1–2 GHz	Long-range surveillance En route traffic control
S	2–4 GHz	Moderate-range surveillance Terminal traffic control Long-range weather
C	4–8 GHz	Long-range tracking Airborne weather detection
X	8–12 GHz	Short-range tracking Missile guidance Mapping, marine radar Airborne intercept
K_u	12–18 GHz	High-resolution mapping Satellite altimetry
K	18–27 GHz	Little use (water vapor)
K_a	27–40 GHz	Very-high-resolution mapping Airport surveillance
Millimeter	40–100+ GHz	Experimental

Figure 1.4 Radar band and usage [1,3]

The main radar functionalities are listed below [4]:

1. *Detection and search*

   These functions give information about the presence of a target within the area under test. The operation of detection is performed through the processing of the received echo. Specifically, it is based on establishing a threshold at the output of the receiver. If the receiver output is large enough to exceed the threshold, a target is said to be present. If the receiver output is not of sufficient amplitude to cross the threshold, only noise is said to be present. This is called threshold detection. It is important to underline that this step is basically to "identify" the target signal from clutter and\or noise.
2. *Tracking*

   This functionality provides the plots and the trajectory of a target by using the return echo signals. The radar can estimate the tracked object's course, the target speed, and the closest point of approach. The radar tracker seeks to determine which plots should be used to update the tracks. The first step consists of updating all of the existing tracks to the actual time in order to predict their new position according to the most recent state estimate (e.g., position, heading, speed, acceleration) and the assumed target motion model (e.g., constant velocity, constant acceleration). The second step is to try to associate the plots to tracks after having updated the estimates.

In order to accomplish this task, a sophisticated set of algorithms like constant false alarm rate are applied. The usage of these sophisticated algorithms makes the system able to track the target also in hostile environments, namely in the presence of nonuniformly distributed clutter or even in the presence of jamming.

3. *Imaging*

   This function aims at generating two dimensional radar images of the observed object or area. These images will be used for several purposes, like detection and classification. Depending on radar available resource, this step will generate well-focused high-resolution range profiles (HRRPs) or two-dimensional inverse synthetic aperture radar (ISAR) images (Figure 1.5) of a moving target. To generate 2D ISAR images, radar pulses in the slow-time domain should be coherently integrated. This means that the radar must track the target for a while. This step is mainly composed of sub-blocks as, for example, image focusing and image formation [5–10]. Some quality indexes may be defined as in [11] to assess the quality of HRRP or ISAR images before using them for classification. Quality index should measure both the quality of the radar image (e.g., in terms of focus degree) and its quality for the purpose of classification (e.g., quality of features that will be used for classification, such as target shape, number of scatterers, etc.).

4. *Classification and recognition [12–16,41,42]*

   The recognition process aims at recognizing the type of a target under test (e.g., a ship rather than an aircraft) or one class of a target from another (a city car rather than a truck). This process makes use of the classification function. The classification consists of process of learning rules or models from training data to generalize the known structure and to classify new data with these rules. In the field of radar application, the classification term refers to a coarse recognition, that is, it is the capability of matching the target to a meta-class like aircraft, wheeled vehicle, and so on.

   Two techniques exist for achieving target classification and recognition:

   - "Model matching technique (MM)." This technique is based on a direct match between the image obtained by received signal and the reference one that belongs to a set of templates [14]. The construction of an exhaustive and consistent database is mandatory. To accomplish this task, the database must contain templates of each kind of targets at all possible geometries. This requirement is needed in order to make the classifier stable with the imaging geometry (e.g., target aspect angle).

     After the database has been built, a target is recognized as that one relative to the template that better matches the reconstructed images. A success measure of the comparison is given by a definition of a match indicator, namely the Match Score [15], that in the most general case is based on the

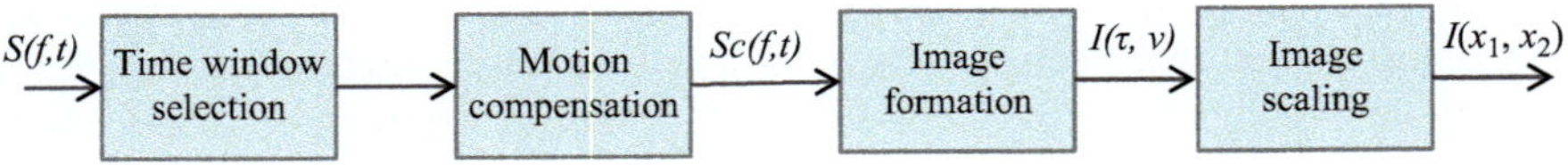

Figure 1.5 ISAR processing scheme [5–10]

correlation coefficient between the two images, the image with the target under test and that one in the database. Although this method is conceptually simple, the main issue is related to the database size that is directly dependent on the amount of the target classes and the potential degrees of freedom.

- "Feature-based technique (FB)" [4,13,16]. This technique overcomes the constraints of the MM technique; in fact the classification is made by using a set of features that are a unique characterization of the target classes and that constitute the features database or feature space. The choice of the features (e.g., geometrical features, polarimetric features, etc.) is an important task to be performed before classification as the classifier performance depends upon the selected features.

A FB classification algorithm is depicted in Figure 1.6.

The FB method consists of comparing the features carried out from the reconstructed image with a set of features from the database. When a fine match exists between the features set from the processed images and those from a particular class within the database, the target is recognized belonging to that class. Therefore, it can be seen that classification divides the feature space into several classes based on a decision rule. From a mathematical point of view, this approach consist of taking an input vector $X = \{X1, X2, \ldots, XN\}$, with N the number of the features, and to assign it to one of M discrete classes, $Cm = \{c1, c2, \ldots\ cM\}$. The input vector, namely feature vector, is used for the learning of the classifier and a labeled object c, namely output vector, specifies the input class. The classifier has to exploit a data set with N labeled examples $D = <X, C>$, namely training data, in order to build a hypothesis that can correctly predict the class label of a new value of the input variable. This results in finding a function defined as which fits the training data in the best way. It is important to underline that when the used model fits the training data too well the overfitting problem may occur. In other words, it may happen that the classifier has memorized the training examples, but it has not learned to generalize to new situations. Conversely, the underfitting occurs when the model does not fit the data well enough as an excessively simple model has been used.

This function in machine learning is known as activation function or discriminant function and the surfaces are the decision surfaces. These regions enclose the input space into decision regions within the feature space. Whenever the discriminants are linear functions, the decision surfaces are hyperplane (Figure 1.7).

Several methods exist for determining the decision surface like Bayesian classifier, support vector machine classifier, and neural network classifier [43].

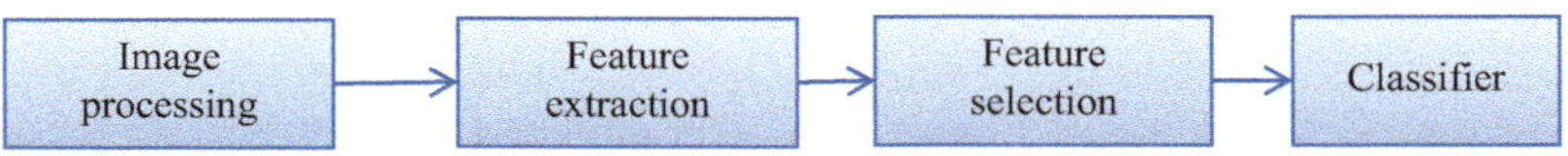

Figure 1.6 High-level architecture of classification processing based on imaging processing

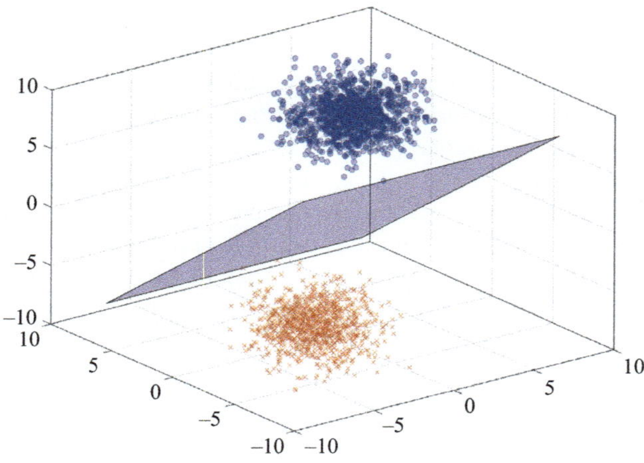

Figure 1.7 Classification of two classes in 3D feature space and decision boundary

5. *Weather observation*

   This process deals with the phenomena of the atmosphere, especially weather conditions, in order to detect, recognize, and measure precipitation event, wind parameters, and other meteorological effects.

These are only some examples of radar functionalities. Summarizing, after the detection operations, the execution of the classification steps may be necessary to identify the target detected. For this purpose, imaging operations may be carried out in order to extract useful information about the target for its classification. This procedure can be used to identify meteorological phenomena and more generally in remote sensing, and in non-cooperative target recognition (NCTR) application.

1.3.2 Radar classes and functionality

As stated in the previous paragraph, particular attention is paid to the radar functions as a way to distinguish the different radars types. Specifically, the radar classes are grouped by each functions defined in Section 1.3.1.

Therefore without loss of generality, the following radar systems can be identified [1,3]:

1. *Detection and search radars*

   Search radars scan a wide area with pulses of short radio waves. These systems usually scan the area 2–4 times a minute with the aim of providing information on the distance and the azimuth of a target. Information on the target's height may or may not be available depending on the antenna features. The angular sector and the radar range rely upon the radar characteristics. For instance, the azimuth could vary from 90° to 360°, while regarding radar range, thousand kilometers can be even reached through a HF over-the-horizon radar.

2. *Tracking radars*

   Tracking radar exploits a scan over a restricted angular sector (e.g., 20°–30°) within a limited range area for detection and tracking of multiple targets.

   Tracking radars include

   - *Single tracker radars [single target tracker (STT)]*

     These radars track a single target at a time at fast data rate giving an accurate tracking of a maneuvering target.
   - *Automatic detection and tracking (ADT) radars*

     These radars track multiple targets (e.g., hundreds/a few thousand targets) simultaneously.
   - *Phased array radar tracking*

     This radar tracks more than one target at high data rate like STT. The beam is electronically switched from one angular position to another in a few microseconds.
   - *Track while scan radars*

     These radars track and search simultaneously over a limited angular sector with a moderate data rate. Equivalent of ADT radars, a number of targets at a time can be treated. It is used to rapidly scan a narrow angular sector both in azimuth and elevation. Scanning can be performed with a single, narrow beamwidth pencil beam or with two orthogonal fan beams (one for the azimuth and the other for elevation one).

3. *Imaging radars*

   Synthetic aperture radar (SAR), ISAR, and side-looking airborne radar (SLAR) are referred to as imaging radars. SAR is employed on airborne or satellite platform. It allows achieving high resolution in cross-range by coherently processing of sequentially received signal of a target or scene over a time of observation. ISAR system is similar to SAR, except that it exploits high resolution in range and the relative motion between the target and the platform to obtain high resolution in Doppler domain and therefore in cross-range domain. SLAR radar provides a high resolution in range and in cross-range by exploiting a narrow beamwidth antenna. The output of the imaging radars is usually a 2D of a target or a scene.
4. *Classification and recognition radars*

   Automatic target recognition and NCTR systems attempt to automatically categorize targets into one of a set of classes.
5. *Weather radars*

   Weather radars, also called weather surveillance radar and Doppler weather radar, can resemble search radars. This radar uses radio waves along with horizontal, dual (horizontal and vertical), or circular polarization. The frequency selection of weather radar is a compromise between precipitation reflectivity and attenuation due to the atmospheric water vapor. The purpose is to locate precipitation, calculate its motion, and estimate its type (rain, snow, hail, etc.). Modern weather radars are mostly pulse-Doppler radars, capable of detecting the motion of rain drops in addition to the intensity of the precipitation. Both types of information can be analyzed to determine the structure of storms and their potential to cause severe weather conditions.

1.3.3 Radar applications

The radar system embraces multidisciplinary applications. The applications are diversified in many fields among which the most important one are listed next:

1. *Military application*
   Military applications involve, for instance, the air defense and battlefield radars.
   - Air defense field comprises all operations devoted to detect aerial targets and to determine their position, course, and speed. The aerial defense space extends up to 500 km per 360° of scan angle.
   - Battlefield includes search, detection, and tracking of unknown moving targets within a restricted sector.
2. *Remote sensing application*
   Remote sensing applications involve weather observation, planetary observation, underground probing, and mapping of sea ice.
3. *Air traffic control application*
   Air traffic control provides services (e.g., air traffic monitoring and regions meteorological mapping closed to the airports and en route) to all private, military, and commercial aircraft operating within its airspace.
4. *Law enforcement and highway safety application*
   This application provides services (e.g., speed measurements, countermeasures actuation in the presence of obstructing or people behind a vehicle or in the blind zone) to enhance safety in road infrastructures.
5. *Aircraft safety and navigation application*
   This application provides services (e.g., mapping of the airport weather in order to classify the regions of precipitation and wind shear, terrain avoidance) to enhance safety in aviation infrastructures.
6. *Ship safety application*
   Ship safety relies upon the radar systems in the field of the surveillance of harbor and river traffic in order to guarantee a certain level of navigation in poor visibility condition and to reduce maritime accidents.
7. *Space application*
   Space application makes use of the radar systems for space vehicle clocking and landing on the moon, for planetary exploration. Moreover, the ground-based segments are used for detection and tracking of satellites, space debris, and for radio astronomy.
8. *HRR applications*
   This application includes security and border surveillance, through the wall and ocean imaging, automotive safety, and medical diagnostics.

   Figure 1.8 depicts some commercial radar systems in function of each mentioned application.

1.3.4 Radar applications and functionalities

For the sake of clarity, Table 1.2 describes how each radar functionality (Section 1.3.1) can be employed in application field (Section 1.3.2).

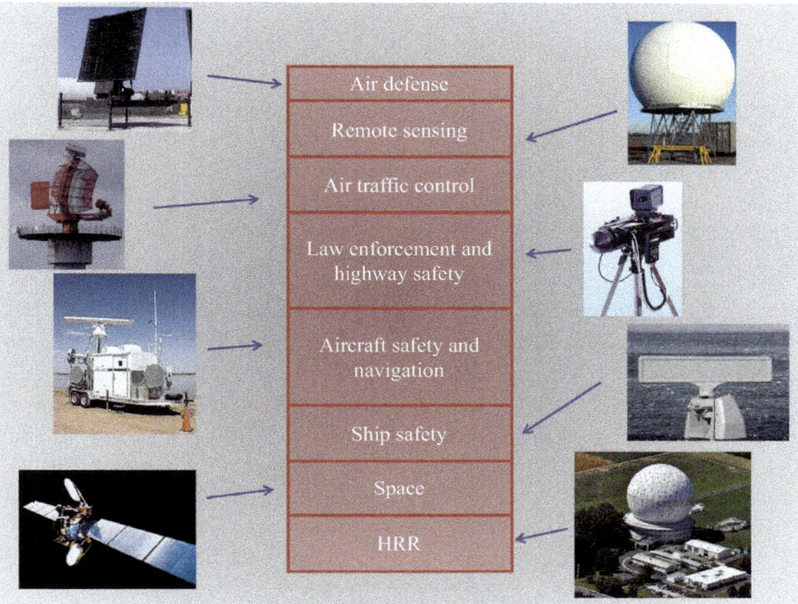

Figure 1.8 Radar system and applications

Table 1.2 Radar functionalities and applications

Radar functionalities	Radar applications
Detection and search	All
Tracking	Military, remote sensing, law enforcement, navigation, ship safety, space
Imaging	Military, remote sensing, law enforcement, navigation, ship safety, space, HRR
Classification and recognition	Military, remote sensing, navigation, space, HRR
Weather	Remote sensing, space, navigation, HRR

1.3.5 Advanced concepts for modern radars

The current research in the microwave radar field is addressing the design and development of radar systems with high performances, robustness, and reliability, low weight, size, and power consumption, high adaptability, and capability to be configured as a network.

There is a wide range of new applications requiring innovative high-performance radar [17], some of the main ones are as follows:

- Spacecraft and/or aircraft (including unmanned aerial vehicles) high-performance radar systems for remote sensing and security in regional areas (sea and river traffic control, micro oil-spill detection, landslide monitoring, coastal erosion, small targets detection on airport surface areas, urban and sub-urban car traffic control, others).

- Detection, classification, and recognition of unknown air, surface, and/or underground targets through high-resolution target imaging up to centimeter resolution and/or exploiting polarimetric signature.
- High detection performance and tracking accuracy of space targets including satellites and debris.

For the above applications, new radars must operate in a wide range of different conditions, which can rapidly change in space and time. In fact geographical location, morphology, urbanization level, as well as the type of target to be detected and tracked strongly differ form case to case. Modern radars should therefore achieve adaptability and the first step in this direction is the capability of creating a very detailed picture of the operational environment that should be maintained updated during operation. In a modern radar system, the effectiveness of the adaptability implementation requires an involvement of both the receiver and transmitter subsystems. This is a fundamental necessity to move toward the challenging idea of realizing cognitive radar systems [18], which means to:

- Achieve an advanced signal processing capable of learning by sensing the environment.
- Exploit the feedbacks coming from the radar receiver to adjust the transmitter parameters in an effective and robust manner.
- Perceive the information associated with the radar returns.

As can be inferred from the previous high-level requirements, learning is the key for realizing cognitive radars. The collection of real data during experimental campaigns is a fundamental way to build up a knowledge about the environment. Learning is then carried out through statistical analysis on the acquired radar data [19,20].

Looking at the application of cognitive radars, one of the main classes of radars that can achieve substantial benefits is the class of multifunction radars [21]. The continuous advances in the area of phased array antenna technology and of computing capabilities are enabling the combination of several functions in the same advanced radar system. The possibility of integrating in a single radar multiple capabilities such as detection, tracking, surveillance, imaging, and weapon guidance can strongly reduce the overall system cost, and at the same time it can improve the radar performance. It is important to observe that multifunction radars need to be multiband in order to cope with multiple needs like achieving long-range coverage, narrow antenna beams formation, wide signal bandwidth, and high immunity to sea and weather clutter. From the technological point of view, implementing a multiband radar means to be able to generate phase stable RF signals in a very flexible way over carriers ranging from low frequency up to extremely HF band [22]. At the receiver side, the same level of reconfigurability and precision is required from the RF front-end and the analogue-to-digital converters (ADCs) [23].

1.3.6 From a single radar unit to radar network topology

As a further development, modern radar systems are demanded to cooperate in a network of radars in order to provide enhanced detection, tracking, and ATR performance. The main advantages of a radar network derive from the higher system resilience and the increased information available due to observation of targets from the multiple different transmitter–receiver pairs.

A multistatic radar system can be defined as one that includes several separated transmitting and receiving stations coupled together for cooperative target observation.

A network of radars is expected to benefit from the spatial separation between radars. The main advantages of a network of radars are:

- A higher detection capability of stealthy and low-observable targets.
- The possibility to use multiple receiver/transmitter pairs that offers the potential for tailored coverage and richer information source to enable more accurate location, high-resolution imaging, and target recognition.

The price to be paid for these advantages is an increase in system complexity and processing. In particular, synchronization and beam pointing are more difficult to implement.

The most important features, strictly connected between them, that characterize a multistatic system are:

- *Degree of spatial coherence*. For spatial coherence, we refer to the capability to maintain a strong phase stability of the equipment in separate stations and consequently to use the phase information of the received signals. A common time and frequency reference signal is necessary in order to establish the synchronization between transmit and receive units, and this synchronization should be maintained during the operation. Of course, the degree of spatial coherence is high in the case of co-located receiver and transmitter as they share the same local oscillator. Conversely, in a radar network since the distance between transmitters and receivers may be large, it may be difficult to establish and maintain this synchronization. This is the main disadvantage of a multistatic system.
- *Information fusion level*. The information fusion level describes the stage at which data between transmitter–receiver pairs is combined. The choice of information integration level will be influenced by the degree of spatial coherence of the radar system itself and the desired application.

In dependence of the level of coherence, multistatic radars can be placed into three main classes:

- *Full coherent systems*: Time, frequency, and phase of transmitted signals are completely synchronized during the entire processing interval. Phase, frequency, and phase synchronization should also be guaranteed between transmitters and receivers.

- *Short-term coherent systems*: Only frequency synchronization is required during the acquisition time. A relatively phase offset between sensors is admitted. In this type of systems, a high relative frequency stability between transmitter–receiver pairs is required, and it is generally achieved by means of synchronization links (radio, optic, fiber, etc.). This level of coherence would be a requirement for each bistatic pair.
- *Incoherent systems*: Only time synchronization between separated stations is necessary. This may be the case where stations nominally operate at the same frequency, but separate oscillators are used, which may, for example, be subjected to different frequency drift. Obviously, any coherent processing algorithms are excluded, and thus phase information is generally removed through envelop detection. These system are simpler than coherent systems. However, the phase information elimination leads to information losses, for example, it is not possible to either exploit coherent detection processing or estimate the Doppler frequency shift.

We can conclude that to maintain a high level of coherence system tends to become more complex and expensive. A good summary of modern synchronization techniques is given in [24].

From the point of view of the information, fusion-level multistatic radars can be placed into two main classes:

- *Centralized processing*: Data fusion takes place prior to any thresholding or decision based on information from any individual transmitter–receiver pair. When the radio signal integration level is used, the raw signals from different nodes are jointly processed. If the video signal integration level is used, the data is combined in much the same way as radio signal fusion, but with phase information discarded before the final data fusion process. This method of fusion looks to make gains in detection through incoherent summation in a similar way to that described for noncoherent pulse integration. Generally in this case a data link with large handling capability is required.
- *Decentralized processing*: Individual detections take place separately for each transmitter–receiver pair. With the plot integration level approach, a preliminary decision regarding the presence or the absence of a target may be made in each node, while a final decision is made for the fusion center as a result of combining the preliminary decisions coming from all nodes. When the track integration level is adopted, not only the detection process but also the tracking processing is applied in each station. The estimated tracks coming from the spatially separated nodes are fused, false tracks eliminated while the true track parameter estimates take advantages from the spatial diversity of the radar network. The required data link handling capacity is reduced drastically when a decentralized integration is used.

In general, we can conclude that when a higher information integration level is used better detection performance is obtained. On the other hand, the system becomes more complex and higher handling capacity is required.

1.4 Issue in current radar networks and future trends

As discussed before, the development trend of radar systems is strongly focused on very specialized applications where extremely high performances are required, such as sub-centimeter-range resolutions [25] or multispectral radar imaging of complex targets [26]. Another fundamental requirement to be fulfilled by the next generation of radars is the reconfigurability, which is needed to realize multi-functional radar systems exploitable over several applications. Flexibility and reconfigurability can be really effective when the radar architecture moves to a full-digital implementation through the concept of software defined radar [27,28].

In current digital radar systems, signal transmission and signal processing are usually carried out at IF or at baseband [1]. Accurate baseband or IF radar signals are generated by direct digital synthesizers (DDSs) up to a few hundred megahertz [1] and up-converted to RF making use of frequency multipliers [29–32], analogue mixers with LO [33], or by controlling a phased locked loop [1]. At the receiver, the signal is down-converted at IF using classical analogue mixers and then digitized with high-performance ADCs [high equivalent number of bits (ENOB), high spurious free dynamic range (SFDR), low power consumption] [30].

Nowadays, DDSs for signal generation up to 6–7 GHz are available [34], but they show poor performance because of low SFDR (about 30 dB) and high phase noise. Typically, direct RF generation is carried out at L-band (1–2 GHz) [35] or S-band (2–4 GHz) [36] with acceptable SFDR (up to 54 dB). For commercial receivers, analogue-to-digital conversion directly at RF can be carried out for low-frequency radar at VHF (30–300 MHz) and P band (250–500 MHz) [37]. Current ADCs would allow a fully digital receiver operating at L-band, but performance would be poor due to low ENOB and SFDR. In fact, commercial high-performance ADCs with high sampling rate up to 5 GSps exist [38] but they exhibit an ENOB equal to 7–8 bit, which is quite low for coherent radars. On the other hand, ADCs with higher performance in terms of ENOB and SFDR (up to 9–10 bit and about 65 dB, respectively) are available at the expense of the sampling rate that can be reduced to 2 GSps [39].

This confirms that fully digital radar receiver up to L-band can be successfully realized; however, the implementation of an RF digital front end at frequencies above 1–2 GHz is unfeasible exploiting the current technology.

In this context, the use of hybrid technologies merging the conventional radar architecture with new generation photonic systems seems the solution for the realization of fully digital radars [23]. In fact, the generation of HF signals is achievable with an electro-optic hybrid approach: the high phase stability of linear electro-optical modulators and of laser sources make these devices suitable for generating HF signals (from C band up the W band) with superior phase stability at the transmitter section of the radar systems, avoiding the problems of conventional radars as, for example, low dynamic range and high phase noise, as mentioned earlier. From the receiver point of view, photonic-based digitalization offers high sampling rates, wide bandwidth together with extremely low jitter with carrier-independent performance and with the possibility to simultaneously deal with multiple signals [40].

1.5 Summary

Radars operating principles have been briefly recalled at the beginning of this chapter, together with the basic radar theory and terminology. The radar main parameters to be taken into account during the design of the sensor have been introduced and a high-level block diagram interconnecting the main radar components has been described. Even if the very first developments of the radar have been carried out in a military domain, afterward this technology moved also toward civil applications and several specialized radar configurations have been proposed. For this purpose, a section of the chapter has been devoted to the definition of the radar functionalities and consequently to the identification of radar classes. As modern applications require increasingly demanding requirements and extremely high performance, a section has been devoted to advanced concepts form modern radars, which should overcome a number of technological gaps and should be able to operate in a rapidly changing environment. These capabilities can be effectively realized through the implementation of cognitive radars, which means systems capable of learning by sensing the environment and of consequently adjust transmitter and receiver parameters. Looking at the application of cognitive radars, one of the main classes of radars that can achieve substantial benefits is the class of multifunction radars. The possibility of integrating in a single radar multiple capabilities such as detection, tracking, surveillance, imaging, and weapon guidance can strongly reduce the overall system cost, and at the same time it can improve the radar performance. Implementing a multiband radar means to be able to generate phase stable RF signals in a very flexible way over carriers ranging from low frequency up to extremely HF band. At the receiver side, the same level of reconfigurability and precision is required from the RF front-end and the ADCs. An effective implementation of flexibility and reconfigurability can be achieved if the radar architecture moves to a full-digital implementation through the concept of software-defined radar. In this context, the use of hybrid technologies merging the conventional radar architecture with new-generation photonic systems seems the most promising solution. As a matter of fact, an electro-optic hybrid approach guarantees very stable signal generation within an extremely wide bandwidth, ranging from C band up to W band. On the receiver side, photonic-based digitalization offers high sampling rates, wide bandwidth, together with extremely low jitter and the possibility to simultaneously deal with multiple signals.

References

[1] M. I. Skolnik, *Radar Handbook*, 3rd edn., New York, USA: McGraw-Hill Companies, 2008.

[2] R.C. Dorf, *The Electronic Engineering Handbook*, 2nd edn., CRC Press, 1997.

[3] D. Jenn, "Radar fundamentals," http://faculty.nps.edu/jenn/Seminars/Radar Fundamentals.pdf [Accessed 30 January 2019].

[4] Federation of American Scientists, "NATO AAP-6-Glossary of terms and definitions," https://fas.org/irp/doddir/other/nato2008.pdf [Accessed 2 July 2015].

[5] M. Martorella, E. Giusti, F. Berizzi, A. Bacci, and E. Dalle Mese, "ISAR based techniques for refocusing non-cooperative targets in SAR images," *Radar, Sonar & Navigation, IET*, vol. 6, no. 5, pp. 332–340, 2012.

[6] M. Martorella and F. Berizzi, "Time windowing for highly focused ISAR image reconstruction," *IEEE Transactions on Aerospace and Electronic Systems*, vol. 41, no. 3, pp. 992–1007, 2005.

[7] S. Brisken, M. Martorella, and J. Worms, "Multistatic image entropy based autofocus," in *Radar Symposium (IRS), 2013 14th International*, 2013.

[8] M. Martorella, F. Berizzi, and B. Haywood, "Contrast maximisation based technique for 2-D ISAR autofocusing," *Radar, Sonar & Navigation, IEE Proceedings* , vol. 152, no. 4, pp. 253–262, 2005.

[9] J. Walker, "Range-Doppler imaging of rotating targets," *IEEE Transaction on Aerospace and Electronic Systems*, vol. 16, no. 1, pp. 23–51, 1980.

[10] E. Giusti and M. Martorella, "Range Doppler and image autofocusing for FMCW inverse synthetic aperture radar," *IEEE Transactions on Aerospace and Electronic Systems*, vol. 47, no. 4, pp. 2807–2823, 2011.

[11] J. Steyn and W. Nel, "Using image quality measures and features to choose good images for classification of ISAR imagery," in *Radar Conference (Radar), 2014 International*, Lille, France, 2014.

[12] D. Blacknell and L. Vignaud, "ATR of ground targets: fundamentals and key challenges," *NATO Lecture Series SET-172*, pp. 1-1–1-32, 2013.

[13] P. Tait, "Automated recognition of air targets: fundamentals and jet engine modulation," in *NATO Lecture Series SET-172*, NATO, 2011, pp. 2-1–2-22.

[14] M. Martorella, E. Giusti, L. Demi, *et al.*, "Target recognition by means of polarimetric ISAR images," *IEEE Transactions on Aerospace and Electronic Systems*, vol. 47, no. 1, pp. 225–239, 2011.

[15] S. Musman, D. Kerr, and C. Bachmann, "Automatic recognition of ISAR ship images," *IEEE Transactions on Aerospace and Electronic Systems*, vol. 32, pp. 1392–1404, 1996.

[16] R. Touzi, "Target scattering decomposition in terms of roll-invariant target parameters," *IEEE Transactions on Geoscience and Remote Sensing*, vol. 47, no. 1, 2007.

[17] M. Richards, J.A. Scheer, and W.A. Holm, Principles of Modern Radar: Basic Principles, New York, USA: SciTech Publishing, 2010.

[18] S. Haykin, "Cognitive radar: a way of the future," *IEEE Signal Processing Magazine*, vol. 23, no. 1, pp. 30–40, 2006.

[19] S. Haykin, R. Bakker, and B.W. Currie, "Uncovering nonlinear dynamics: the case study of sea clutter," *Proceedings of the IEEE*, vol. 90, no. 5, pp. 860–881, 2002.

[20] M. Greco and F. Gini, "X-band sea clutter non-stationarity: the influence of long waves," in S. Haykin, Ed., *Adaptive Radar: Toward the Development of Cognitive Radar*. Hoboken, NJ: Wiley, 2006, ch. 5.

[21] S.L.C. Miranda, C.J. Baker, K.D. Woodbridge, and H.D. Griffiths, "Knowledge-based resource management for multifunction radar," *IEEE Signal Processing Magazine*, vol. 23, no. 1, pp. 66–76, 2006.
[22] F. Laghezza, F. Berizzi, A. Capria, *et al.*, "Reconfigurable radar transmitter based on photonic microwave signal generation," *International Journal of Microwave and Wireless Technologies*, vol. 3, pp. 383–389, 2011.
[23] P. Ghelfi, F. Laghezza, F. Scotti, *et al.*, "A fully photonics-based coherent radar system," *Nature*, vol. 507, pp. 341–345, 2014.
[24] M. Weib, "Synchronisation of bistatic radar systems," in *IGARSS 2004. 2004 IEEE International Geoscience and Remote Sensing Symposium, 2004*, pp. 1750–1753, vol. 3.
[25] K. B. Cooper, R. J. Dengler, N. Llombart, B. Thomas, G. Chattopadhyay, and P. H. Siegel, "THz imaging radar for standoff personnel screening," *IEEE Transactions on Terahertz Science and Technology*, vol. 1, no. 1, pp. 169–182, 2011.
[26] P. van Dorp, R. Ebeling, and A. G. Huizing, "High resolution radar imaging using coherent multiband processing techniques," in 2010 IEEE Radar Conference, Washington, DC, 2010, pp. 981–986.
[27] B. L. Cheong, R. Palmer, Y. Zhang, M. Yeary and T.Y. Yu, "A software-defined radar platform for waveform design," in *2012 IEEE Radar Conference*, Atlanta, GA, 2012, pp. 0591–0595.
[28] T. Debatty, "Software defined RADAR a state of the art," in *2010 2nd International Workshop on Cognitive Information Processing*, Elba, 2010, pp. 253–257.
[29] F. Yang, X.-H. Tang, and T. Wu., "The scheme and key components design of W-band coherent doppler velocity radar front-end," in *ASIC, 2007. ASICON '07. 7th International Conference on*, 22–25 October 2007, Hangzhou, China, pp. 356–359.
[30] P., Slawomir, "FMCW radar transmitter based on DDS synthesis," *in Microwaves, Radar & Wireless Communications, 2006. MIKON 2006. International Conference on*, 22–24 May 2006, Krakow, Poland, pp. 1179–1183.
[31] C. Wagner, A. Stelzer, and H. Jager, "A 77-GHz radar transmitter with parallelised noise shaping DDS," in *Radar Conference, 2006. EuRAD 2006. 3rd European*, 13–15 September 2006, Manchester, UK, pp. 335–338.
[32] T.E. Derham, S. Doughty, K. Woodbridge, and C.J. Baker, "Design and evaluation of a low-cost multistatic netted radar system," *Radar, Sonar & Navigation, IET*, vol. 1, no. 5, pp. 362–368, 2007.
[33] L. Zhai, Y. Jiang, X. Ling, and W. Gao, "DDS-driven PLL frequency synthesizer for X-band radar signal simulation, " in *ISSCAA 2006, International Symposium on System and Control in Aerospace and Austinautics*, 19–21 January, pp. 344–346.
[34] V.Y. Vu, A.B. Delai, and L. Le Cloirec, "Digital and super-resolution ultra wide band inter-vehicle localisation system," in *Communications and*

Electronics, 2006. ICCE '06. First International Conference on, 10–11 October 2006, Hanoi, Vietnam, pp. 446–450.

[35] S.E. Turner and D.E. Kotecki, "Direct digital synthesizer with ROM-less architecture at 13 GHz clock frequency in InP DHBT technology," *IEEE Microwave Components Letter*, vol. 16, no. 5, pp. 296–298, 2006.

[36] B. Ferguson, S. Mosel, W. Brodie-Tyrrell, M. Trinkle, and D. Gray, "Characterisation of an L-band digital noise radar," in *Radar Systems, 2007 IET International Conference on*, 15–18 October, 2007, Edinburgh, UK, pp. 1–5.

[37] C.J. Peacock, G.S. Pearson, and W.N. Dawber, "Wideband direct RF digitisation using high order Nyquist sampling," in *Waveform Diversity & Digital Radar Conference - Day 2: From Active Modules to Digital Radar*, 2008 IET, 9–9 December, 2008, London, UK, pp. 1–6.

[38] R.L. Thompson, E.L.H. Amundsen, T.M. Schaefer. P.J. Riemer, M.J. Degerstrom, and B.K. Gilbert, "Design and test methodology for an analog-to digital converter multichip module for experimental all digital radar receiver operating at 2 Gigasamples/s, *IEEE Transactions on Advanced Packaging*, vol. 22, no. 4, pp. 649–664, 2006.

[39] E2V EV10AQ190 – 5 Gsps for one channel – Data sheet.

[40] G.C. Valley, "Photonic analog-to-digital converters." *Optics Express*, vol. 5, no. 15(5), pp. 1955–1982, 2007.

[41] T. Cooke, M. Martorella, B. Haywood, and D. Gibbins, "Use of 3D ship-scatterer model from ISAR image sequences for target recognition," *Elsevier Digital Signal Processing*, vol. 16, pp. 523–532, 2006.

[42] M. Martorella, E. Giusti, A. Capria, F. Berizzi, and B. Bates, "Automatic target recognition by means of polarimetric ISAR images and neural networks," *IEEE Transactions on Geoscience and Remote Sensing*, vol. 47, no. 11, pp. 3786–3794, 2009.

[43] C.M. Bishop, *Pattern Recognition and Machine Learning*, Berlin, Germany: Springer, 2007.

Chapter 2
Electronic warfare systems and their current issues

Maurizio Gemma[1], *Antonio Tafuto*[1], *Marco Bartocci*[1] *and Daniel Onori*[2]

2.1 Chapter organization and key points

This chapter reviews the concepts and current implementations of the electronic warfare (EW) systems. Following a logical categorization, the systems realizing electronic support, attack, and protection will be taken into account, highlighting for each of them the peculiar strengths as well as the open issues.

2.2 Electronic warfare scenario

EW systems aim at controlling and using the electromagnetic spectrum to sense possible threats, impede enemy assaults via the spectrum, and secure the communications [1]. These platforms operate to negate or mitigate the efficacy of the enemy threats present in a hostile scenario constituted by radar/navigation systems, infrared sensors, as well as telecommunications equipment [2].

For example, a typical goal of an EW apparatus is to detect the signals produced by a radar threat and to re-radiate back engineered interfering signals to deceive, mask, or alter the real echo signal, and to provide wrong information to the tracking algorithms [1,2] so that:

- the enemy cannot detect and locate the friendly platforms;
- the enemy cannot launch the terminal threats (e.g., missiles); and
- if a terminal threat is successfully launched, its automatic guidance is corrupted by unacceptable errors.

Generally speaking, in EW systems three main tasks can be identified:

- *Electronic support (ES)* actions collect the data necessary to detect and identify potential hostile sources in a crowded scenario.

[1]Elettronica SpA, Rome, Italy
[2]Institut National de la Recherche Scientifique (INRS) – Energie, Matériaux et Télécommunications, Montréal, Québec, Canada

- *Electronic attack (EA)*, previously known as electronic counter measures (ECMs), consists in neutralizing enemy electronic systems by emitting tailored electromagnetic signals toward them.
- *Electronic protection (EP)*, previously known as electronic counter–counter measures (ECCMs), protects military forces against any electromagnetic threat (e.g., enemy and friendly EA).

The demanding features provided by EW devices require a high-performance transmit/receive apparatus (transceiver), where the major components are

- a wideband antenna or antenna array, with beam steering capabilities;
- a receiver subsystem able to provide digital output messages, called pulse descriptor words, with the information of frequency, amplitude, pulse width, etc., of the detected threat;
- a transmitter able to produce radio frequency (RF) signals with the purpose of block or deceive the systems generating the enemy electromagnetic threats; and
- a processor that performs gathering, analyzing, and recognizing of threats.

In this chapter, we will describe the main architectures and techniques for implementing the above-mentioned tasks of ES, attack, and protection.

It is useful here to describe also how a generic threat is organized. From this point of view, an offensive action against a platform can be split into the following steps:

- Detection, identification, and location of the platform.
- Launch of the terminal threat against the platform. This can be accomplished by a fire control system or by a missile system. The missiles can be launched by different platforms such as ships, submarines, and aircraft;
- Tracking of the terminal threats toward the target. The terminal threats are generally guided by an inertial navigation system, with or without mid-course corrections.

In the terminal phase, the missile seeker must accurately track the target in order to generate the correct information necessary to home on the targeted platform. Therefore, in order to be sure to hit the target, during the last part of their course, the missiles switch on a seeker for tracking the target and steering toward it. For this purpose, the most widely used seekers are of the RF type due to the fact they are "all weather." The seekers of the electro-optic type, instead, have problems in case of fog or smoke, and they are generally used as back-up systems in case RF main seeker fails to operate due to EA actions.

It is also important to underline here that the generic military RF systems include both radar-type and communication-type emitters [1–3]. The possible dangerousness of radar-type emitters is self-evident as they are mounted on surveillance and weapon systems. On the other hand, today's military operations are also characterized by the need for robust and secure communications networks that can transfer multimedia information in real time to fixed and mobile stations, thus providing all the elements of the network with a capillary fast command chain and an updated complete situation awareness. Therefore, communication-type RF emitters are placed side by side to military radar systems and are considered as a possible threat by the EW apparatuses.

2.3 ES receivers

ES receivers are systems capable of detecting, classifying, identifying, and locating the RF threats generated by radar, surveillance, and weapons systems, and by the communication apparatuses of the military equipment present in the environment [1,4].

The ES sensors devoted to the radar emitter detection and those devoted to the communications analysis have considerable different requirements because of the different operative frequencies of the waveforms exploited in the two applications. In fact, radar signals operate at high frequencies (>1 GHz) and are generally based on periodic short pulses. Therefore, the main functions performed by the wideband ES radar sensor have to be executed in a very short time before a rapid weapon system can become a threat [1,2,5].

On the other hand, in communications emitters the waveforms are generally based on spread-spectrum signals, and ES communication receivers require longer acquisition and analysis in order to decrypt the hidden information. In the past, the communications-type emitters typically worked at low carrier frequencies (usually <1 GHz) with respect to the radar type. However, due to the increasing frequencies of the communication threats, nowadays ES receivers for radar- and communications-type emitters share very similar architectures.

Moreover, thanks to the considerable advances of digital technologies, the current trend in EW sensors is the use of digital receivers, exploiting their simple reconfigurability to adapt to both the scenarios of radar and communication signals.

2.3.1 ES receivers for radar emitters

Typically, the radar-type emitters of surveillance and weapons systems operate at high frequencies (usually >1 GHz) to exploit directive antennas for target direction measurements. They use phase- or frequency-modulated waveforms, either pulsed or continuous wave (CW), to exploit pulse compression techniques to increase the accuracy in the measurements of the target distance. The carrier frequency of the signal waveforms can be changed following specific patterns, even from pulse to pulse, within the so-called frequency agility bandwidth, in order to provide robustness against enemy countermeasures.

Given the aforementioned scenario, ES radar receivers can be classified based on their application as follows:

- *Radar warning receiver (RWRs).* A RWR warns of the presence of threat emitters and identifies their direction of arrival (DOA) within the spatial and frequency coverage of the sensor.

- *Electronic support measurement (ESM) receiver*. An ESM detects DOA and classifies all the radar emitters within the sensor spatial and frequency coverage.
- *Electronic intelligence (ELINT) receiver*. An ELINT provides prolonged and accurate measurements of all the characteristics of a radar emitter (frequency and waveform patterns) in order to collect the data necessary for the analysis and modeling of the associated threats. These data are used to update the emitter libraries of the two previous sensor types (RWR and ESM) for the correct identification of the threats.

These ES architectures and their operational needs are discussed in detail in Section 2.3.4.

2.3.2 ES receivers for communications emitters

Nowadays, military tasks require robust and secure communications networks that can transfer multimedia data in real time among fixed and mobile stations, thus providing all the nodes of the network with a full and updated awareness of the scenario.

In the past, the communications-type emitters typically worked at lower frequencies (usually <1 GHz) with respect to the radar type. However, today they are moving toward higher frequencies to use wider signal bandwidths and enhance the number of the communication channels.

The communications threats are virtually spread all over the RF spectrum. They usually exploit longer RF waveforms with respect to the radar emitters, with modulated amplitude phase or frequency, seldom in the form of pulse bursts. Spread-spectrum and/or frequency hopping strategies are usually used, which are the communications equivalent of the radar pulse compression and frequency agility techniques, as they are both exploited as anti-interference or anti-jamming techniques.

Given the aforementioned scenario, ES communications intercept systems can be distinguished as follows:

- Communications electronic support measures (CESM) apparatuses, if they have to capture RF signals, detect DOA, and classify all the communication emitters.
- Communication intelligence (COMINT) systems, if they have to examine the threat signals over an extended period of time for providing intelligent classification of the emission characteristics and scenario by prolonged and accurate measurements of all the characteristics of a communication emitter.

The tasks of CESM systems are the search, interception, identification, and classification of communications signals in the surrounding scenario in order to enable possible EW countermeasures, such as location, analysis, or jamming of the relevant network nodes. However, these tasks are often challenging due to multiple reasons. First, the presence of several friendly or civilian communication devices. Second, because modern military communications equipment adopts advanced techniques for low probability of intercept (LPI) operations, such as the direct digital synthesis of spread spectrum encrypted coded signals and intelligent frequency hopping.

Concerning the COMINT equipment, its main task is to record the communication signals from threat emitters for the successive decryption of the hidden information, and for the analysis of their LPI features so as to devise the most appropriate jamming countermeasures.

2.3.3 Basic ES sensors architectures

Typically, wideband receiver are defined to instantly cover the I- (8–10 GHz) or J-band (10–20 GHz) or both of them, whereas wide-open devices operate in the complete RF band of radar threat ranging from the D- to J- or K-band (1–40 GHz). Usually, wideband receivers are also supported by an instantaneous frequency measurement (IFM) receiver, which is employed to estimate the carrier frequency of the detected signal. Alternatively, narrow-band detectors sense a narrow portion of the spectrum with high sensitivity and can generally sweep in the D- to J-band (1–20 GHz) [4]. Given these definitions, five common configurations of ES receivers are listed in the following:

- crystal video receiver (CVR), either wide-open or wideband channelized, that is, based on multiple wideband sensors;
- superheterodyne receiver (SHR), either swept narrow-band or swept wideband;
- channelized receiver (CHR), where the required wideband frequency coverage is divided into a large number of narrow-band SHR channels with high sensitivity and dynamic range;
- transform receivers (TRs), such as Bragg cells, microscan, or compressive receivers; and
- cued receivers, which are hybrid configurations of the previous architectures.

A comparison of the main features of these configurations is reported in Table 2.1. In the following, the different configurations are described in detail.

Table 2.1 ES receiver comparison table

Parameters	**Receiver type**				
	CVR	**SHR**	**CHR**	**TR**	**Cued**
Instantaneous B_{RF}	Excellent	Poor to fair	Good to excellent	Good	Excellent
Sensitivity	Poor to fair	Good to excellent	Excellent	Good	Excellent
Simultaneous signal handling	Poor	Poor to fair	Good	Good	Good
Dynamic range	Poor to good	Excellent	Good	Good	Excellent
Signal parameter measurement accuracy	Fair	Good	Good	Good	Excellent
Configuration complexity and cost	Low to medium	Medium to high	High to very high	Medium to high	High

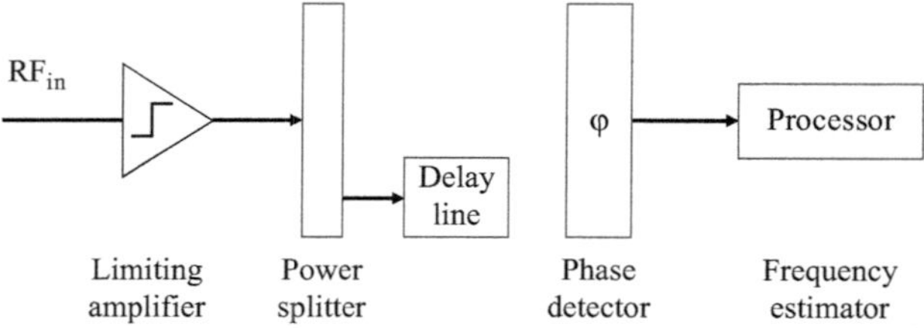

Figure 2.1 IFM receiver functional block diagram

2.3.3.1 IFM receiver

The IFM receivers have been broadly adopted in EW systems of the past generation and are still applied nowadays in current equipment, although the future trend is to replace them with a digital receiver (as discussed in Section 2.3.5).

The principle of operation of the IFM receiver is shown Figure 2.1. The RF input signal $s_i(t) = A\cos(2\pi ft)$, where f is the instantaneous frequency carrier, is amplified by a limiting amplifier, and then split in two equal amplitude halves. One of the halves is delayed with respect to the other through a delay line of length L, which provides a time delay $\tau = L/c_p$ (c_p is the electromagnetic wave propagation speed in the delay line device). The two signals are then applied to a phase detector [4]. This way, the phase of the two halves is compared and the phase shift $\varphi = 2\pi f\tau$ is proportional to the carrier frequency. Because τ is a constant, the angle φ is a measure of the signal carrier frequency f. For a given delay, the unambiguous band covered is the one for which the phase change is $1/\tau$. For example, for an unambiguous bandwidth equal to 1 GHz, the required delay is 1 ns.

The most pressing problem in applying the IFM is caused by simultaneous signals. When more than one signal is present at the same instant, the detected phase shift is not a meaningful measure of the frequency of any of them.

2.3.3.2 Crystal video receiver

CVRs are the most used solutions in the past generation of EW receivers. They are primarily exploited in RWR because they cost-effectively cover widebands: I- or J-band, or both, or even the complete RF band of possible radar threats ranging from the D- to the J- or K-band (in this case they are defined as wide-open receivers).

The modern generation of CVRs shows both a good sensitivity, because of the addition of an amplifier in front of the detector, and the capability of accurately measure the carrier frequency due to the addition of an IFM receiver. Current implementations are used as wide-open CVR for RWR application and wideband CVR for medium-class ESM equipment (see Section 2.3.4). The basic block diagram of a CVR is shown in Figure 2.2, which is essentially a square-law envelope detector followed by a video amplifier. A low-noise amplifier can be used after the antenna to enhance the sensitivity [1,2].

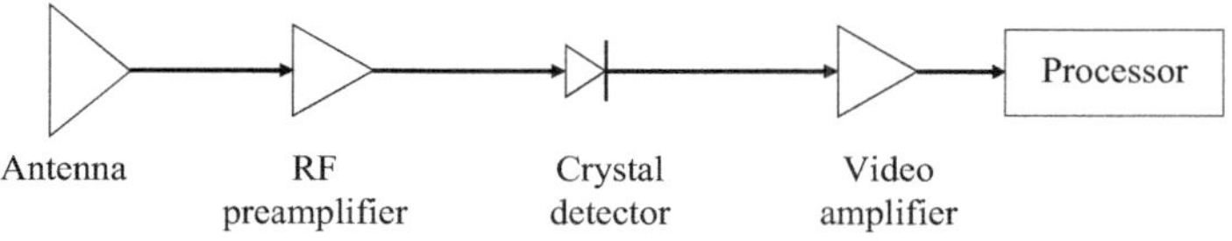

Figure 2.2 CVR functional block diagram

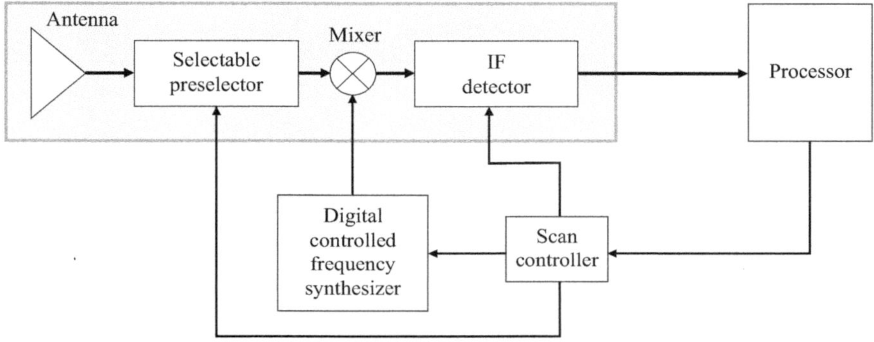

Figure 2.3 Basic block diagram of an SHR

The CVR sensor implementation is generally exploited to provide the following signal (pulse) instantaneous parameters:

- pulse amplitude
- pulse width
- DOA
- time of arrival (TOA)
- frequency.

Because the CVR is small, usually it is installed very close to the antenna in the different channels exploited for DOA.

2.3.3.3 Swept superheterodyne receiver

The SHR is the most widely adopted receiver in demanding radar and communications applications due to its high sensitivity, dynamic range, and frequency selectivity. In EW applications, narrow-band SHRs (with instantaneous bandwidth < 1 GHz) are used primarily in ELINT applications (see Section 2.3.4) in order to isolate an emitter signal from the environment, measure its information in detail, and suppress any undesired interfering signals. On the contrary, wideband (instantaneous bandwidth > 1 GHz) scanning SHRs are used as part of ESM equipment (see Section 2.3.4) to detect and discover threat signals in harsh, highly dense scenarios because their selectivity strongly reduces the received pulse density [1].

A basic block diagram of an SHR is depicted in Figure 2.3. The principle of operation stems from mixing the incoming input with a tunable local oscillator (LO) in order to translate it to an intermediate frequency (IF). A preselector is required before mixing to reject the image signal [1,2]. Modern SHRs use a software-enabled preselected scanning strategy, suited to the current operative scenario, over a number of bands within the RF interval covered by the system (the bandwidths may also be adjacent so as to completely cover the required RF range). Obviously, the dwell time in a detected band can be set longer if a high threat density required detailed investigation, while can be shortened in bands where there are few threats.

SHRs are characterized by a higher sensitivity than the CVRs due to the reduced instantaneous RF bandwidth that implies less integrated noise. However, since SHRs multiplex in time the observed bandwidths, they have limitations in the obtainable probability of intercept (POI) unless appropriate scan dynamic programming is applied. Generally, the POI with SHRs is lower than the one with CVRs.

2.3.3.4 Channelized receiver

The CHR architecture covers the required RF bandwidth through a number of parallel contiguous channels, each of them having an operative bandwidth that approximately matches the expected threat emitter bandwidth, which is considered between 200 MHz and a few GHz in modern implementations [4]. The basic scheme of principle is depicted in Figure 2.4. Each channel is able to provide all of the instantaneous pulse signal parameters. However, this architecture is very

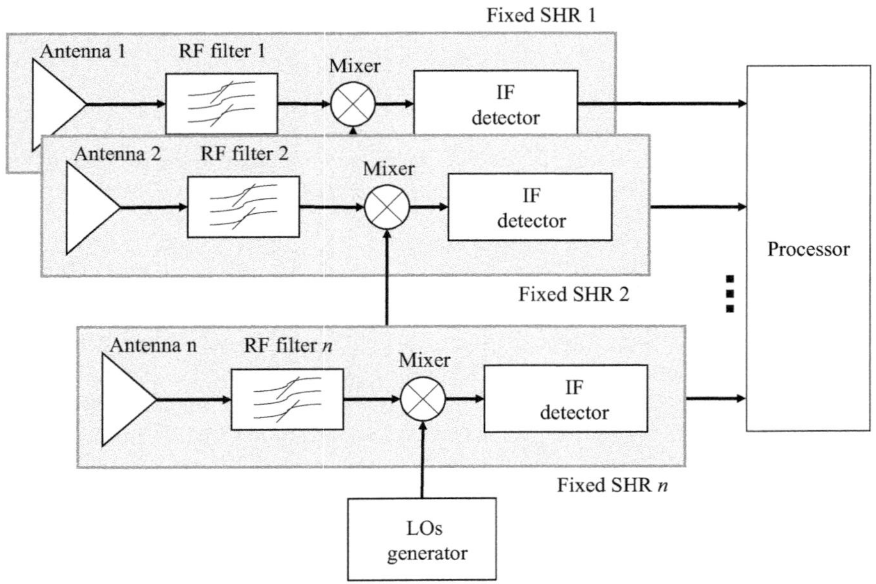

Figure 2.4 Schematic block diagram of a channelized receiver

complex, bulky, and expensive; thus, it has been used only on ground-based and large transport aircraft ELINT equipment [1].

The important advantage of this architecture is its ability to handle multiple simultaneous threat signals that overlap in time (as is the case in highly dense RF band segments). This solves the issue of poor performance of IFM sensors. In fact, two simultaneous signals usually fall in separate high-dynamic channels and can be accurately processed. An accurate processing can occur even if the two signals fall in two adjacent channels, thanks to the very sharp bandwidth roll-off of each channel. Moreover, even if the signals are acquired within the same channel, the digital filtering and processing implemented in modern CHRs is capable of accurately extracting the desired information [4,5].

Practical configurations of the channelized architecture (in terms of reduced complexity and volume) are utilizing the properties of a narrowband SHR, while also providing wide frequency coverage, by using a large number of parallel channels, each one tuned to a slightly different RF center frequency.

2.3.3.5 Transform intercept receivers

In past implementations, the cost and complexity of wideband channelizers led to the development of acousto-optic receivers. This type of receiver, which was implemented with various technologies, approximates the operation of a Fourier transform in hardware [1]. The most important technology is the Bragg cell receiver, which exploits the acousto-optical characteristic of some materials (e.g., a crystal of lithium niobate, $LiNbO_3$) when fed by an RF signal, to produce a deflection of a laser beam as it passes through the material and provides a measure of the RF signal frequency.

In these crystals, an incident beam of coherent light is generated by a laser source. The amplitude of the RF signal is determined by the level of the light striking an array of photodetectors, as shown in Figure 2.5. Because multiple signals produce multiple output deflections, conceptually, this receiver is a replacement for the CHR.

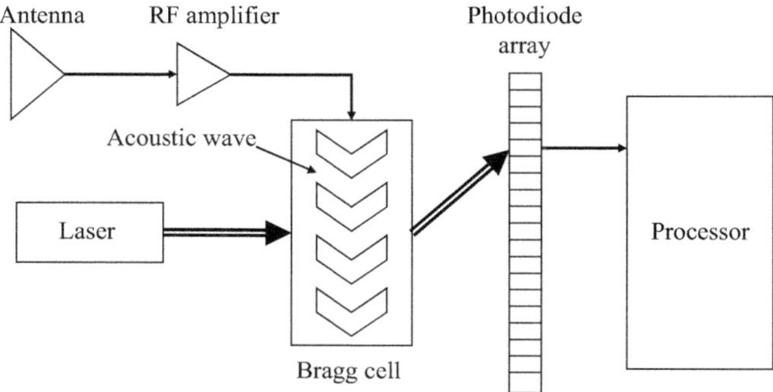

Figure 2.5 Schematic block diagram of a Brag cell receiver

2.3.4 ES receivers implementations and requirements

Currently, the electromagnetic scenario that an EW apparatus needs to face is characterized by a large number of emitters with several different functionalities. In order for the EW system to effectively analyze and communicate the actual threats, it is essential to have EW receivers capable of detecting several simultaneous signals that may span over a wide range of frequencies, from D- to K-band (1–40 GHz). As fundamental features, such an EW receiver should be able to provide:

- the capability to handle multiple-frequency threats simultaneously;
- a near-to-unity POI for each threat;
- the capability of measuring the frequency of the threat with adequate resolution; and
- the capability of single-pulse acquisition and parameter measurement.

As introduced in Section 2.3.1, EW sensors can be classified based on their application as RWRs, ESM, and ELINT receivers. In those receiver implementations, the architectures introduced in Section 2.3.3 are exploited to reach the required functionalities. The implementation and operational needs of RWRs, ESM receivers, and ELINT receivers are detailed in the following sections.

2.3.4.1 Radar warning receivers

RWRs are fundamental ES passive surveillance systems used in applications that demand high POI but do not require high sensitivity. Their main task is to provide a warning to the pilot of an aircraft if a threat emitter is illuminating the aircraft. In more detail, RWRs provide the following functions:

- detection of the threatening emitter;
- coarse measurement of the DOA of the threat emission, with respect to the longitudinal axes of the platform, provided that several directional antennas and associated receivers are available on the platform;
- measurement of the fundamental parameters of the threat signal (e.g., amplitude, pulsewidth, DOA, TOA);
- identification and sorting of the threat signals when the signal environment is not significantly dense; and
- classification of the threat signals by comparison with properly prepared data libraries.

The typical schematic block diagram of a wide-open RWR (i.e., complete RF coverage of the required threat emitter spectrum), which is the most commonly used RWR architecture, is shown in Figure 2.6. The broad RF band is filtered into parallel channels. Then, the crystal video device detects the envelope of the received signal, but removes any information related to the frequency and phase of the emitter waveform. The video filter then performs a further smoothing, enabling anyway the detection of the shortest pulses. The parallel filtered RF signals, before detection, are also sent to an IFM device (see Section 2.3.3), which provides the signal frequency estimation.

The RWR operational requirements are collected in Table 2.2 [1,4].

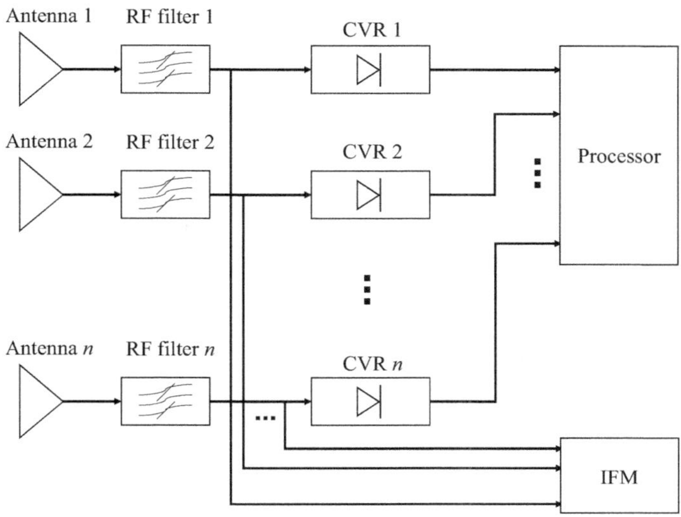

Figure 2.6 Schematic block diagram of a wide-open RWR

Table 2.2 Required performance for a RWR

RF coverage	2–18 GHz (with IFM) 28–40 GHz (without frequency measurement)
Video bandwidth	20 MHz
Sensitivity	–45 dBm typ. (without preamplifier) –60 dBm typ. (with preamplifier)
Instantaneous dynamic range	40 dB
Total dynamic range	70 dB (with preamplifier)
Frequency measurement accuracy	2–10 MHz with harmonic IFM
Minimum pulse width	200 ns
Maximum pulse width	Continuous wave (CW)
Maximum no damage input power	1 W CW; 100 W pulsed duty 1%
Settling time	<1 μs

2.3.4.2 ESM receivers

A fundamental mission of an EW system is the surveillance and reconnaissance (S&R), which is performed through the ESM receivers and several sensors either active or passive. S&R, also known as scouting, is a basic mission involving the location, identification, and tracking of the objects within the monitored environment. It represents a mandatory task prior all the other defense activities of the superior platform (e.g., a ship, a helicopter, a platoon, etc.).

In particular, the S&R activities of the considered EW system do not require an immediate reaction to radar emissions, that is, the instantaneous coverage of the whole operational frequency band. This implies that the frequency coverage can be performed with a time division scanning (e.g., by a super-heterodyne receiver).

Besides the capability of threat warning, ESM receivers require a very high sensitivity to detect every threat emitter in a very dense scenario. In these systems, a CHR or a combination of CHR plus wide-open crystal video receiver architecture can be used (Figure 2.7).

The ESM receiver operational requirements are collected in Table 2.3.

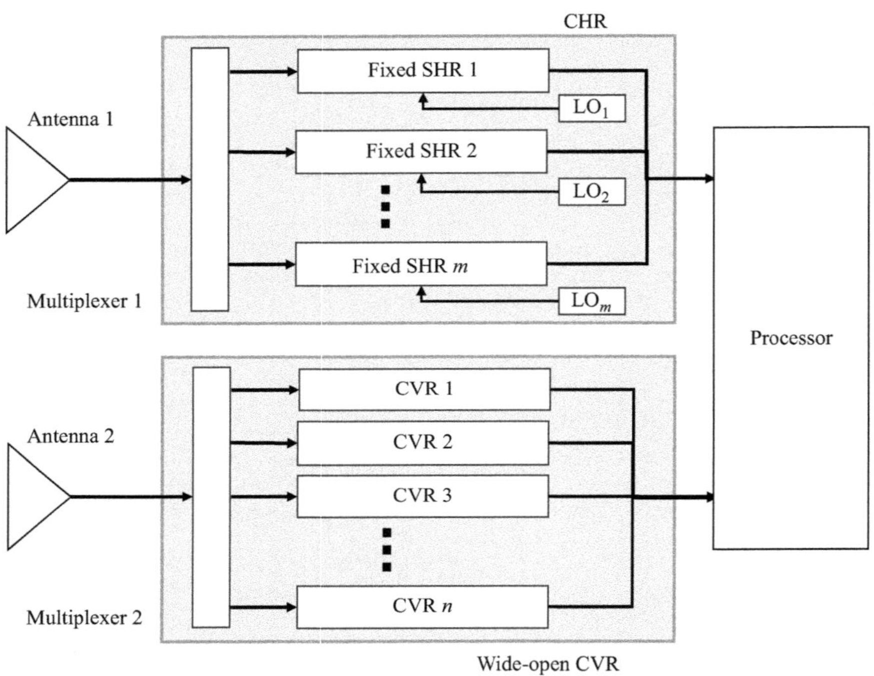

Figure 2.7 Schematic block diagram of an ESM receiver

Table 2.3 Required performance of an ESM receiver

RF coverage	0.5–18 GHz, future trends: 0.5–40 GHz
Instantaneous bandwidth	1 GHz
Frequency resolution	1 MHz
Instantaneous dynamic range	>40 dB
Spurious-free dynamic range	>50 dB
Noise figure	<5 dB
Minimum pulse width	100 ns
Maximum pulse width	Continuous wave
Sensitivity	–80 dBm
Maximum Input power	0 dBm

2.3.4.3 ELINT architecture and requirements

ELINT is the activity of information gathering by use of electronic sensors. Its primary focus lies on noncommunications signals intelligence, particularly radar signals. ELINT differs from the surveillance activity in that it encompasses a technical analysis of the received signals that is carried out typically offline since the main focus is not real-time tactical situational awareness (i.e., the Electronic Order of Battle), but rather an in-depth analysis of the received waveforms to classify the potential emitters, build, or update EW radar libraries.

The ELINT receiver presents extremely high sensitivity and selectivity in order to detect the threat signals also transmitted from the lateral lobes of their antennas. ELINT receivers are exploited both for fast and long operations, with acquisition times that span from few microseconds to milliseconds, and are capable of collecting adequate amount of information to provide identification and classification to support RWR and ESM systems operations.

The ELINT architecture, as shown in Figure 2.8, is generally based on an SHR scheme that exploits a number of selectable narrow-band IF filters [1]. These filters permit to achieve the very high sensitivity required by the ELINT applications. If filters with different bandwidth are also used for the analysis of intrapulse and interpulse characteristics of the radar threat signals, a sort of fingerprint of the emitter is useful to provide its identification and classification.

The ELINT operational requirements are collected in Table 2.4 [1,4].

2.3.5 *Digital receiver*

The latest available digital technologies have enabled the implementation of a common architecture for the EW sensors for both radar- and communications-type emitters, thus allowing the so-called EW spectrum sensor [1].

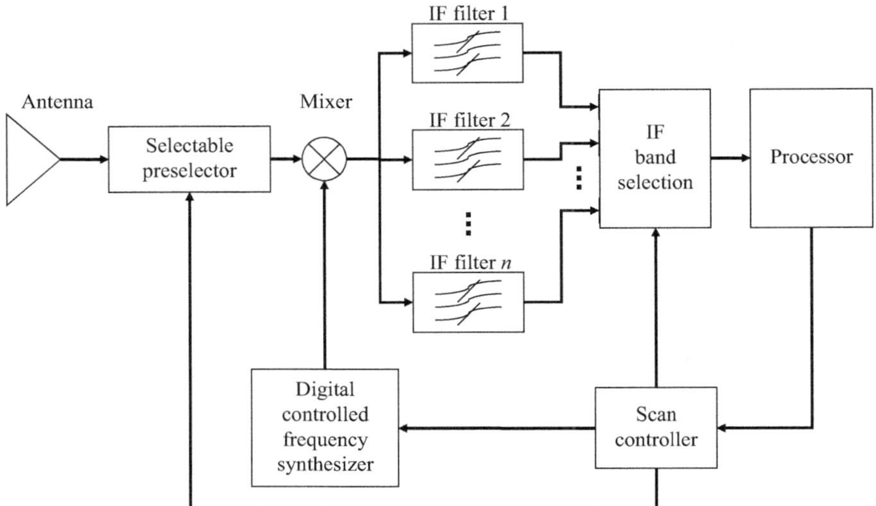

Figure 2.8 Schematic block diagram of a ELINT receiver

Table 2.4 Required performance of a ELINT receiver

RF coverage	0.5–18 GHz, future trends: 0.5–40 GHz
Instantaneous bandwidth	1 GHz
Frequency resolution	1 kHz
Instantaneous dynamic range	>70 dB
Spurious-free dynamic range	>50 dB
Noise figure	<5 dB
Minimum pulse width	Few ns
Maximum pulse width	Continuous wave
Sensitivity	–85 dBm
Maximum input power	0 dBm

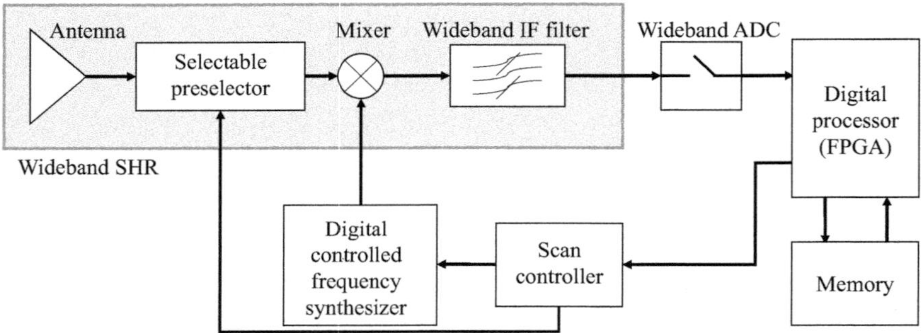

Figure 2.9 Schematic block diagram of a digital receiver

The tremendous evolution of electronic high-speed digital technologies in terms of sample and hold (S/H) circuitry, analog-to-digital converter (ADC), and digital signal processor (DSP) has produced serious changes in the architecture of RF sensors, and particularly in EW ones. These changes have provided better equipment performance and efficiency, reducing volume and power consumption [5,6]. In the case of EW sensors, the novel device at the base of the new architecture is the digital receiver [1,7]. A typical block diagram is shown in Figure 2.9.

The basic structure of a digital receiver is composed of the following parts:

- A wideband SHR stage that performs the down-conversion from RF to IF.
- A wideband ADC, which provides discrete amplitude levels at discrete time instants. The wideband IF filter of the SHR stage also prevents alias distortion after acquisition.
- A processing engine implemented on an application-specific integrated circuit or field programmable gate array (FPGA). The FPGA has the advantage of being reprogrammable according not only to application needs but also to mission needs.
- A storage memory for signal samples recording. This option allows offline software processing to be performed by a processor not hosted on the digital receiver.

To allow multichannel operation, the following functions are required to perform differential measures of phase, amplitude, and TOA:

- A synchronizer that aligns the samples and the processing between different digital receiver channels.
- Some I/O devices for exchanging data.

This digital receiver is able to implement advanced software-defined functionalities in order to guarantee high performance and flexibility at the same time. However, reducing also its size, weight, and power consumption (SWaP) is not a trivial task, primarily due to the encumbrance of the 2–18 GHz (or even 0.5–40 GHz) filter bank and tunable LO (the digital controlled frequency synthesizer in Figure 2.9). In Section 2.5, possible solutions and future developments will be presented.

2.4 EA architectures

The aim of EA systems is to neutralize enemy EW systems, thus limiting their capabilities to attack and defend (in fact, these systems were previously known as ECM). This task is pursued by means of advanced technique that attempt to confuse, deceive, or even blind enemy systems. Therefore, EA equipment attempts to avoid a large number of casualties and damages in missions where the modern scenario consists of advanced weapons systems with high striking capabilities.

EA/ECM transmitting systems cover the same electromagnetic spectrum of ES equipment and are classified in accordance to the type and function of the RF threat system for which they are designed. Therefore, we can find communications ECMs (CECMs) that are designed to block RF communication systems, radar ECMs (RECMs) that take care of radar system threats (often simply called ECM), and infrared ECMs that fight IR systems. In particular, typical applications of RECM systems are for self-protection against radar seekers and missile threats, while CECMs are primarily used for the denial of service in mobile platforms.

With respect to the ECM definition used in the past, the modern EA definition adds the following functions:

- direct energy weapons
- antiradiation missiles
- electromagnetic and nuclear electromagnetic pulses

Focusing on RECM, these systems consist of RF transmitters that generate jamming targeted to deceive and blind enemy radars [1]. Jamming acts by sending an interference signal to the receiver of the threat radar to deceive of corrupt its detection task. Usually, the interference signal is either composed of random noise or is a modified replica of the radar waveform. In current applications, radar menaces cover the wide frequency range 2–18 GHz (with a trend to extend the range to 0.5–40 GHz [2]) and make use of increasingly complex signals and advanced EP skills, becoming more and more robust to jamming attacks [1,2,8]. Consequently, modern EA apparatus has to present advanced RF jammer technology implementing a fully adaptive threat response [1,8], while also aiming at reducing their SWaP as much as possible.

Adaptability and flexibility can be achieved through a software-defined approach that implements digital techniques to sharpen up the system capabilities. Typical examples are the digital RF memory (DRFM) jammers, the most advanced EA self-protection technology today, capable of digitizing and storing the received radar signal threat, then generating and retransmitting an altered version designed to spoof the enemy.

A DRFM device consists of a wideband receiver, a DSP unit, and a wideband transmitter. The block diagram of a DRFM is shown in Figure 2.10. The DRFM receive path is very similar to a conventional SHR with regard to the rejection of spurious signals. The RF input signal is band-pass filtered and amplified in order to suppress the undesired components. Then, the filtered signal is down-converted to baseband by superheterodyning with a LO. A low-pass filter (LPF) is used for anti-aliasing and for removing the intermodulation products generated by the mixer. At the output of the LPF, the signal is sampled and digitized by an ADC with suitable analog bandwidth and precision (i.e., number of bits). The higher the number of bits corresponds to a higher dynamic range of the ADC, but is associated to a slower conversion process. Although the digital side is continuously and quickly evolving, the analog transmit/receive hardware still relies on a stackable implementation due to the lack of tunability in state-of-the-art electronic and microwave technology [9], and great effort is being carried on to reduce SWaP also in EA systems, especially when attempting to cover the 0.5–40 GHz operative range.

2.5 EP architectures

EP is a part of EW systems that include a variety of practices aiming at reducing or eliminating the effects of electronic attacks (EA/ECM) on electronic sensors aboard vehicles, ships, aircraft, or weapons such as missiles. EP is also known as ECCM. In practice, EP often means resistance to jamming.

Modern surveillance radars, in addition to advanced signal processing techniques implemented by a digital architecture (as discussed in Section 2.3.5), employ a number of ECCM techniques, implemented at hardware level, to reduce the effects of the hostile ECM systems. The most important ones are listed below and discussed in the following subsections, while more details can be found in [1] and [4]:

- frequency and pulse-repetition interval agility
- ultralow sidelobes
- multiple sidelobe cancellers (SLCs)
- sidelobe blanker

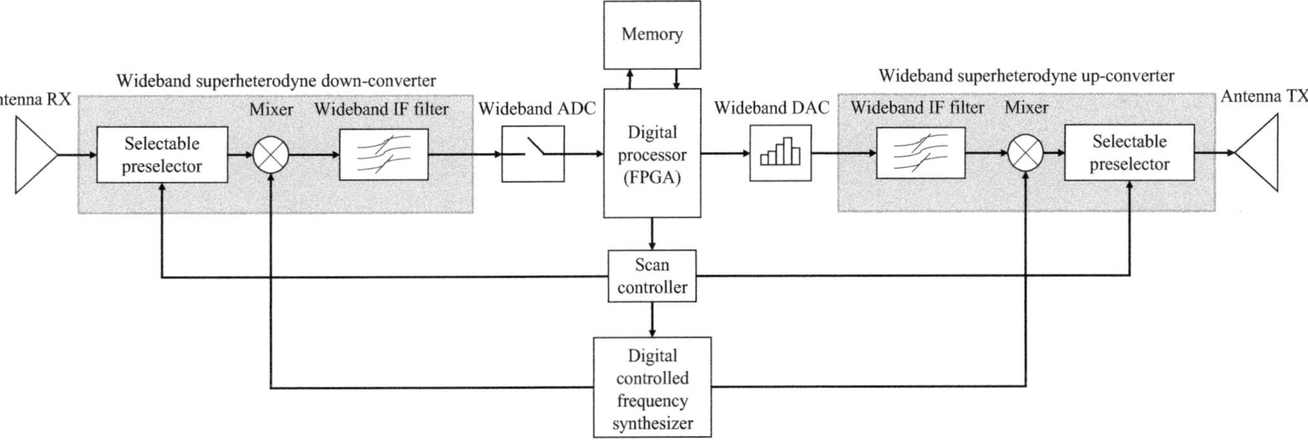

Figure 2.10 Block diagram of a DRFM

2.5.1 Frequency and pulse repetition interval agility

Pulse-to-pulse frequency agility of the generated radar signals can be an effective technique against RECM jammers. The radar frequency agility bandwidth in current systems is usually about 10% of the central frequency, which is enough for spreading the power of the jamming noise signal and reducing its interference. Moreover, false targets produced by RECM enemy systems (e.g., false radar echoes retransmitted by DRFM systems) can occur only at distances greater than the real distance of the jammer [1].

The pulse repetition interval agility has the same effect as frequency agility with respect to the deceptive false targets.

2.5.2 Ultralow sidelobes

Designing antennas for EW systems with reduced sidelobe levels with respect to the main lobe is considered an effective EP strategy. In fact, it strongly reduces the noise jamming power injected from the sidelobes of the antenna to the ES receiver. This way, enemy systems require RECM emitters with extremely high directivity (or effective radiated power) to create false targets detected by the protected EW equipment.

2.5.3 Multiple SLCs

Another EP technique related to an effective design of the antenna is the SLC architecture. It consists of a programmable and adaptive array antenna system capable of reducing the signal jamming power received in its sidelobes. In detail, the antenna system can reshape its radiation lobes in order to position a null of the emission pattern in the direction of the jammers. A common implementation is based on a linear array of elementary dipoles, in which the phase shift (and the gain as well, in more advanced solutions) of the feeding signal is adaptively controlled [1].

2.5.4 Sidelobe blanker

The sidelobe blanker (SLB) consists of an additional receiving channel connected to an omni-directional antenna. This EP system acts deleting the effects to the main receiver produced by strong target reflections and jamming attacks that could be detected through the sidelobes of the antenna. The schematic block diagram of an SLB is shown in Figure 2.11(a). The principle of operation stems from an adaptive management of the gain of the additional receiving channel. This way, any threat signal coming from the sidelobe region of the main antenna is also received by the additional channel with a convenient amplitude and can be compared to the same signal received by the sidelobes of the main antenna (see the radiation patterns of the main and additional antenna depicted in Figure 2.11(b)). Only signals detected by the main lobe of the main antenna can proceed to the radar detection chain.

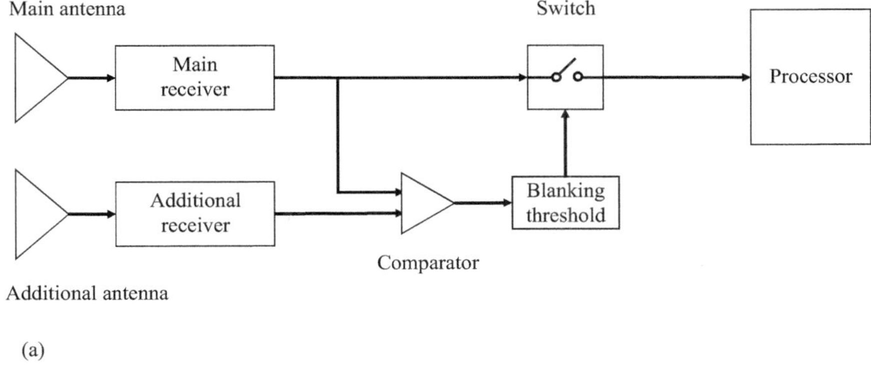

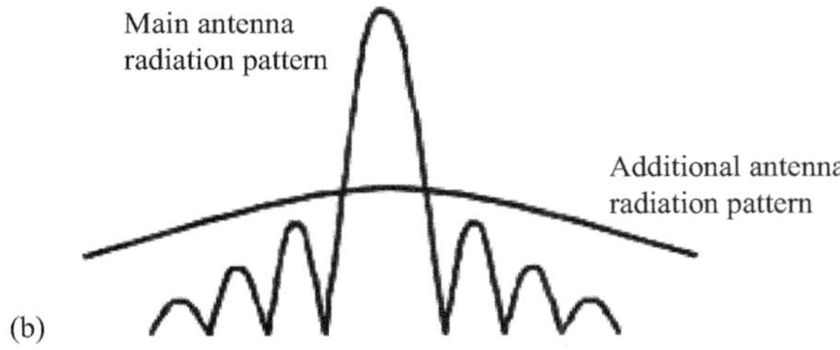

Figure 2.11 (a) Schematic block diagram of the SLB principle; (b) radiation patterns of the main and additional antennas, respectively

2.6 Future developments

One of the main required advancements that the EW systems need is the reduction of the SWaP. The main bottleneck in achieving this reduction is the current state-of-the-art implementation of the uniform filter bank and of the tunability of the LOs. Different proposals have been investigated so far to solve this issue by avoiding the use of those components.

A first solution consists of the digital direct sampling receiver (DDSR), which acquires the RF input with a wideband ADC avoiding any RF down-conversion stage (i.e., without any mixer and LO). In fact, with respect to the digital receiver presented in Section 2.3.5, the DDSR exploits a wideband wide-open ADC, which provides discrete amplitude levels at discrete time instants in the 0.5–40 GHz frequency band.

The schematic block diagram of such a system is depicted in Figure 2.12.

However, conventional ADCs are fundamentally limited by the timing jitter of the sampling source: the higher the sampling frequency, the more detrimental the sampling jitter on the ADC resolution. Therefore, this limit requires a trade-off between ADC bandwidth and resolution. As a result, DDSR's result is practical

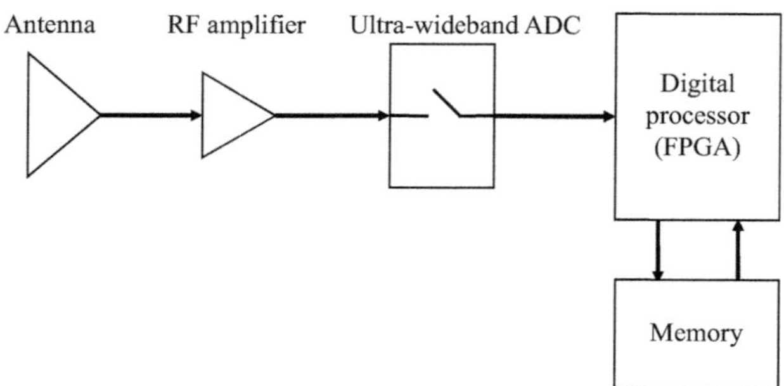

Figure 2.12 Schematic block diagram of a DDSR

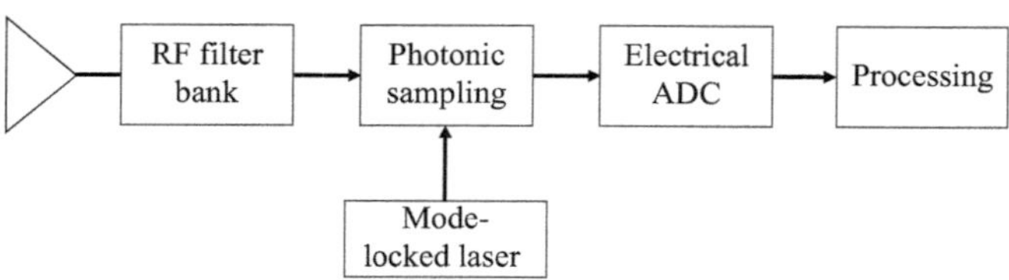

Figure 2.13 Schematic block diagram of a photonic sampling receiver

only at L- and S-band and RF systems are typically designed with narrow-bandwidth acquisition channels [10]. These engineering constraints becomes serious limitations when dealing with broadband signals as ultra-short pulses.

Photonic techniques has been very recently introduced in radar and EW systems for reducing the SWaP and improving the performance [11,12]. In the last few years, in fact, photonic technologies have demonstrated attractive features for microwave applications, as ultra-wide bandwidth, tunable filtering, photonics-based microwave mixing with very high port-to-port isolation, and intrinsic immunity to electromagnetic interferences [13]. Moreover, photonic integration is rapidly growing, producing photonic integrated circuits more and more robust, reliable, compact, and at lower cost.

The photonics-based analog-to-digital technique aims at realizing an ADC with a photonic sampling source (e.g., a mode-locked laser) that directly digitizes the entire operative bandwidth instantaneously and coherently, with a greater resolution compared to a conventional electronic system (Figure 2.13) [11]. Thanks to the extremely low timing jitter of the laser source and to the elimination of the RF tuner, this technique has the potential to demonstrate a signal fidelity more than 100 times better than the state-of-the-art digital electronics. The possibility of implementing the photonics-based ADC exploiting photonic integration also promises to reduce the system encumbrance. Nevertheless, the RF filter bank continues to be essential in most applications where the rejection of image signal is mandatory. In those situations, the final SWaP is unavoidably compromised.

A possible solution has been presented with the time-interleaved photonic sampling, where multiple high precision, low sampling rate electronic ADCs are

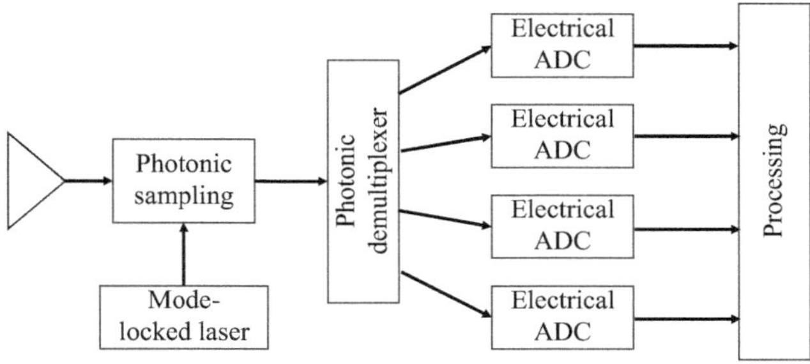

Figure 2.14 Schematic block diagram of a time-interleaved photonic sampling receiver

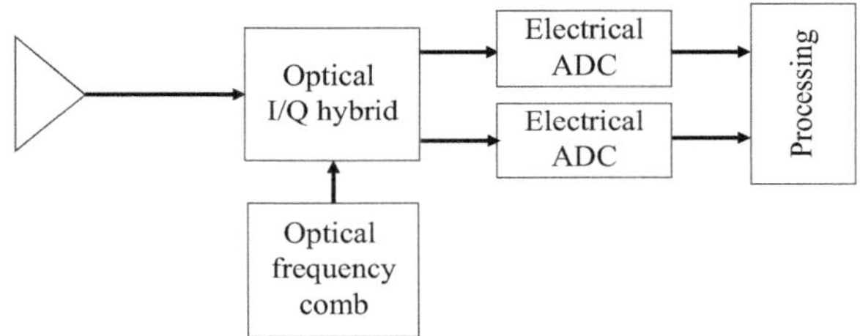

Figure 2.15 Schematic block diagram of a photonic direct conversion scanning receiver

exploited perform a full Nyquist rate sampling of the RF input, as depicted in Figure 2.14.

Another possibility to reduce the SWaP could be using a single wideband scanning receiver that covers the 0.5–40 GHz frequencies by means of a single-stage direct down-converter and a single ADC. In fact, the direct conversion approach would avoid the use of any RF image rejection filter and reduce the final SWaP. Unfortunately, current RF components such as hybrid in phase/quadrature (I/Q) mixers (direct conversion approach) and electronic oscillators lack in tunability and performance stability throughout the required bandwidth [7].

Once again, the use of photonics has been proposed for effectively implementing a broadband scanning direct conversion receiver [12]. This is based on the use of an optical I/Q hybrid coupler and of an optical frequency comb (which is used as a multiple LO source). The scheme is depicted in Figure 2.15, and in its first prototype implementation, it has demonstrated extremely promising performance.

2.7 Conclusions

In this chapter, the principles of operation, adopted technologies and techniques, and operational needs of modern EW equipment have been described. We have shown how they cope with the challenges provided by the evolution of the transmitted waveforms in radar and communications equipment, as illustrated in Chapter 1. Moreover, the effort in improving the performance of these technologies, while reducing their encumbrance and power dissipation, is underlined throughout the chapter.

Future possible developments are also drawn, merging integrated photonics with the traditional EW architectures to reach the next requirements in terms of functionalities, performance, and SWaP reduction.

References

[1] A. De Martino, "Introduction to Modern EW Systems." Norwood, MA, USA: Artech House Publishers, 2012.

[2] R.G. Wiley, "ELINT: The Interception and Analysis of Radar Signals." Norwood, MA, USA: Artech House Radar Library, 2006.

[3] A. Graham, "Communications, Radar and Electronic Warfare." ISBN: 978-0-470-68871-7, 2010.

[4] C. Dantea, "Modern Communications Receiver Design and Technology." Norwood, MA, USA: Artech House Publishers, 2010.

[5] J.B. Tsui, "Microwave Receivers with Electronic Warfare Applications." ISBN: 1891121405, 2005.

[6] M. Golio, and J. Golio, "RF and Microwave Applications and Systems." Boca Raton, FL, USA: CRC Press, 2007.

[7] J.B. Tsui, "Digital Techniques for Wideband Receivers." Raleigh, NC, USA: SciTech Publishing, 2000.

[8] I. Arasaratnam, S. Haykin, T. Kirubarajan, and F.A. Dilkes, "Tracking the mode of operation of multi-function radars." 2006 IEEE Conference on Radar, Verona, NY, USA. DOI: 0.1109/RADAR.2006.1631804.

[9] M. Frater and M.J. Ryan, "Electronic Warfare for the Digitized Battlefield." Norwood, MA, USA: Artech House, 2001.

[10] P. Delos, "A Review of Wideband RF Receiver Architecture Options." Analog Device Tech-Note, 2016.

[11] P. Ghelfi, F. Laghezza, F. Scotti, *et al.*, "Photonics in radar systems: RF integration for state-of-the-art functionality." IEEE Microwave Magazine, v. 16, n. 8, pp. 74–83, 2015.

[12] D. Onori, F. Laghezza, F. Scotti, *et al.*, "A DC Offset-Free Ultra-Wideband Direct Conversion Receiver based on Photonics." EuRAD/EuMC03-02, EuRAD 2016, London.

[13] J. Capmany, and D. Novak, "Microwave photonics combines two worlds." Nature Photonics, v. 1, pp. 319–330, 2007. DOI:10.1038/nphoton.2007.89.

Chapter 3
Microwave photonic concepts and functionalities

Massimiliano Dispenza[1], Luigi Pierno[1], Paolo Ghelfi[2] and Antonella Bogoni[2,3]

3.1 Chapter organization and key points

In this chapter, we introduce the principal concepts and functionalities that photonics can bring in the microwave systems. The discussion will focus in particular on the application of photonics to radars and to electronic warfare (EW) systems. The analysis will compare the performance of several standard microwave subsystems with the latest results obtained by using photonics, either coming from the research labs or, where available, from the pioneering microwave photonics industry. The potentials of photonics in improving the performance with respect to the standard approaches are stated. Moreover, several novel peculiar possibilities that photonics can enable are highlighted.

In order to be fruitful to the largest technical audience, the analysis will avoid discussing the photonic solutions in deep details, which would require the experience of a photonic specialist to be fully understood. Nevertheless, the descriptions will let the reader getting the main concepts, and the reported numbers will allow a direct comparison with the standard microwave system counterparts. For those who will be willing to reach a deeper insight, several references are provided.

After an introductory paragraph on the general benefits brought about by photonics, this chapter takes into account the basic functionalities of microwave systems: the generation and detection of the microwave signal, its transmission and distribution, the signal conditioning for the beamforming function in phased array antennas (PAAs), and the signal filtering. The potentials coming from the integration of photonics are also discussed in a dedicated paragraph. Finally, the main points raised in this analysis are summarized.

[1]Leonardo SpA, Italy
[2]National Laboratory of Photonic Networks and Technology (PNTLab), Consorzio Nazionale Interuniversitario per le Telecomunicazioni (CNIT), Italy
[3]Institute of Technologies for Communication, Information and Perception (TeCIP), Sant'Anna School of Advanced Studies, Italy

3.2 Microwave photonic solutions for future radar and EW systems

Up to now, radar systems have been usually designed for implementing a single specific function as, for example, naval traffic monitoring or meteorological studies. This fixed-function approach has been supported by the design of specific electronics aiming at optimizing the radar performance exclusively in its foreseen application.

Recently, a strong request for multifunctional and multiband sensing systems is raising, aiming at reducing the cost of the radar system by leveraging on the concept of reconfigurability [1]. Moreover, the increasing need of resolution (e.g., in the automotive sector) is also pushing toward high carrier frequencies to increase the available signal bandwidth (see also Chapter 1).

These new radar systems should therefore be capable of adjusting their parameters (e.g., the bandwidth, or the pulse repetition interval, or the central frequency) to adapt to several operational situations: from detection to tracking, even to wireless communication! Consequently, the multifunctional radar transmitters require reconfigurable and software-defined radio frequency (RF) signal generators capable of producing wideband waveforms over carriers ranging up to the millimeter waveband (MMW, above 30 GHz), while maintaining the phase stability necessary for coherent pulse-Doppler processing, target imaging, and clutter rejection [2,3]. At the same time, the multifunctional radar receivers need to digitally acquire the detected signals directly at their frequency range, without any fixed down-conversion, to ensure the system reconfigurability and reliability [4].

Because of these demanding requirements, current electronic technologies are facing serious difficulties in guaranteeing the needed performance. In fact, as will be detailed in the next paragraphs, the direct generation of modulated RF signals by means of direct digital synthesizers with acceptable stability is limited to a few gigahertz, and the necessary process of multiple active up-conversions worsens the generated signal phase noise [2,3]. Similarly, the precision of analog-to-digital converters (ADCs) drops with increasing input bandwidth and sampling speed [5], thus requiring multiple down-conversion stages.

Similar issues are being faced in EW systems. As the radars and the wireless communications are becoming more intelligent and flexible, and are exploiting higher carrier frequencies, detecting and identifying them is requiring increasing efforts. The spectrum to observe is enlarging and the required precision is increasing, which translates into larger and heavier EW systems, equipped with more ADCs and related electronics. This can be done only at the expenses of larger size, weight, and power consumption (SWaP) of the EW system. Unfortunately, this novel must be traded off with the request of system installation on board of requiring platforms as the unmanned vehicles, which show stringent specifications on acceptable payload.

Over the last decade, the use of photonics has been investigated to overcome the performance of electronics in RF systems, giving birth to the new research field named microwave photonics. In fact, photonics shows several properties that are

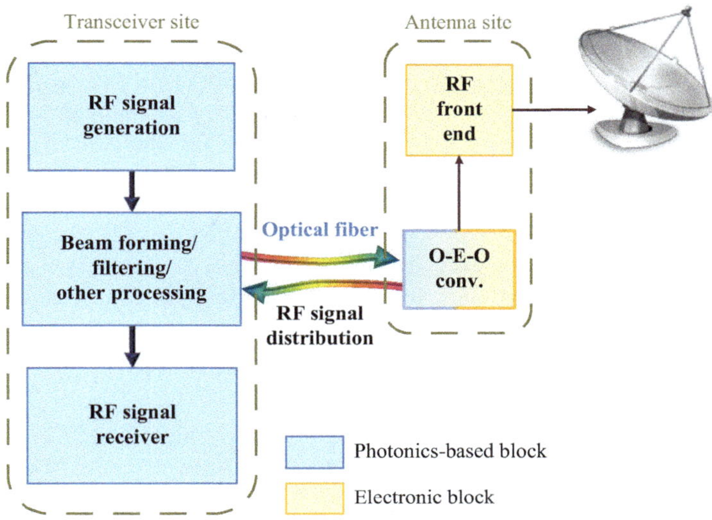

Figure 3.1 Possible exploitation of microwave photonics in a radar or electronic warfare system

fundamental for developing the future microwave systems, and in particular the future radar systems and EW systems [6–9], as sketched in Figure 3.1, which are characterized by a similar general structure:

- Low-phase noise: Lasers characterized by extremely low noise (amplitude or phase noise, summarized in the parameter of the laser linewidth) are now available, and they can be used to generate or detect RF signals with excellent stability, in particular in terms of phase.
- Wide bandwidth: The technological advancements of optical communications have provided photonic devices that allow loading very broadband signals on a laser carrier, and these devices can now be used in microwave photonics as well to manage RF signals with tens of gigahertz of bandwidth, and with central frequencies up to the millimeter waves.
- Easy tunability: Lasers and other devices as the optical filters are easily tunable, and this can be used in microwave photonic solutions to enable an unprecedented frequency flexibility in RF systems.
- Low-loss and low-distortion propagation: These are fundamental properties of optical fibers, and among the reasons for their worldwide adoption for communications. These features can be borrowed by RF systems as well to transport microwave signals across long distances, avoiding the loss and bandwidth problems of the commonly used waveguides.
- Immunity to electromagnetic interferences (EMIs): This peculiar feature of the optical transmission can be clearly of uppermost importance in RF systems operating in a harsh environment.

As will be explained in more details in this chapter, photonics can be used for generating RF signals in a broad range of carrier frequencies with high phase stability, also allowing the simultaneous generation of multiple RF carriers. It enables processing wideband signals, for example, implementing the beamforming in PAAs, or the tunable RF filtering. At the receive side, the photonics-based ADC can guarantee large input bandwidth, high sampling rates, extremely low jitter, and the capability to simultaneously receive multiple signals. Finally, the use of photonics in a microwave system allows the distribution of RF signals through optical fibers between separated subsystems as, for example, a radar transceiver or an EW system and their antennas, with low loss, low distortions, and EMI immunity.

3.3 Photonics-based RF generation and up-conversion

Figure 3.2(a) shows the scheme of a conventional generation of RF signals. It is based on multiple electronic up-conversion stages, each one including an electronic local oscillator (LO) followed by a bandpass filter (BPF) to select the up-converted signal and suppressing the image frequency. With the reported scheme, a modulation signal generated at baseband (BB) or at intermediate frequency (IF) can be

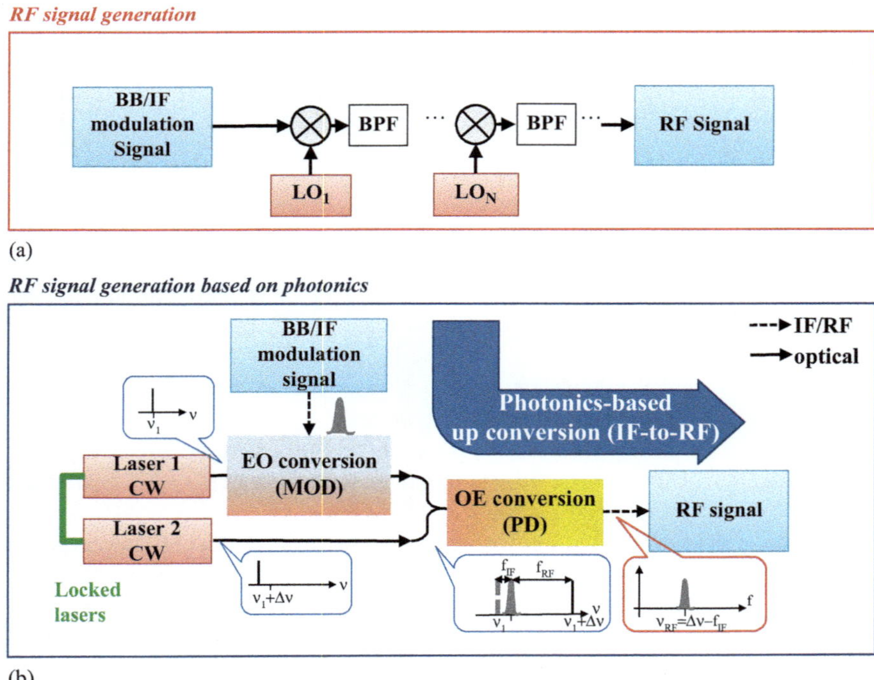

Figure 3.2 (a) Conventional RF generation scheme; (b) photonics-based RF generation

up-converted to the desired RF central frequency. Unfortunately, each of the up-conversion stages introduces non-negligible phase and amplitude noise due to the phase drifts of the LO and the nonlinearities of the active electronic mixers. It must also be underlined that each stage makes use of an LO and a BPF at specific frequencies, and that the different LOs are usually incoherent with each other. Moreover, the phase stability of the LOs decreases as the RF increases.

Figure 3.2(b) shows the general scheme of a photonic RF generation. It is based on the concept of heterodyning, which means detecting two laser signals in a photodiode (PD), thus generating an RF signal proportional to the square of the input optical fields. Let us assume that the two lasers (respectively, *Laser 1* and *Laser 2*) emit ideal continuous wave (CW) optical signals, with electromagnetic fields that can be expressed as:

$$S_1(t) = A_1 \cos(2\pi\nu_1 t + \phi_1) \tag{3.1}$$

$$S_2(t) = A_2 \cos(2\pi\nu_2 t + \phi_2) \tag{3.2}$$

where A_i is the amplitude of the field (with $I \in \{1,2\}$ identifying the laser), ν_i is the optical frequency of the laser, and ϕ_i is the phase of the optical field. When they are coupled together and received by a PD, the PD generates an electrical current (the RF signal) that can be expressed as:

$$S_{\mathrm{RF}}(t) \sim [S_1(t) + S_2(t)]^2 \tag{3.3}$$

After substituting the expressions for $S_1(t)$ and $S_2(t)$, and neglecting the direct-current components and the components at twice the optical frequency (which are too fast to be detected by the PD), the formula above becomes:

$$S_{\mathrm{RF}}(t) \sim A_{\mathrm{RF}} \cos[2\pi(\nu_2 - \nu_1)t + (\phi_2 - \phi_1)] = A_{\mathrm{RF}} \cos\left[2\pi\nu_{\mathrm{RF}} t + \Delta_\phi\right] \tag{3.4}$$

Therefore, an RF signal is generated whose frequency ν_{RF} is equal to the detuning between the two lasers. Thanks to the wide opto-electronic bandwidth of the currently available PDs, the generation of RF signals up to and above 100 GHz can be easily achieved.

If one of the lasers is modulated, for example, in an external optical modulator (MOD) driven by a modulation signal at BB or at IF, as is the case for *Laser 1* sketched in Figure 3.2(b), the heterodyning transfers the modulation to the RF carrier frequency given by the lasers detuning. For instance, if we assume that *Laser 1* is modulated in its amplitude, the expression of the RF signal from the PD becomes:

$$S_{\mathrm{RF}}(t) \sim A_{\mathrm{RF}}(t) \cos\left[2\pi\nu_{\mathrm{RF}} t + \Delta_\phi\right] \tag{3.5}$$

where the amplitude term $A_{\mathrm{RF}}(t)$ holds the BB or IF modulation information originally loaded on *Laser 1*. Since the optical communications field has brought to the market optical modulators with electro-optical bandwidth above 40 GHz (and much larger modulation bandwidths are available from the research labs), photonics allows the generation of ultra-broad bandwidth RF signals with carrier frequencies up to the MMW. In more details, the signal bandwidth is given by the available modulation bandwidth of the optical modulation, while the central RF frequency directly depends on the detuning between the two lasers, and is limited by the available bandwidth of the heterodyning PD.

One of the most important parameters in the generation of RF signals is the phase stability, in particular for surveillance radars and for communications. In the photonics-based scheme of Figure 3.2(b), the phase stability of the generated RF signal depends on the reciprocal phase behavior of the two beating optical signals, as can be seen from the term Δ_ϕ in the expression of (3.4). From the equation it is evident that, if the two lasers have ideally no phase noise (i.e., their phase term ϕ_i is a constant), then the optically generated RF signal has no phase noise as well (the phase term Δ_ϕ is a constant too). Unfortunately, lasers always do have a certain amount of phase noise, and two independent beating lasers would generate an RF signal with a phase term equal to the difference of the lasers phase noise processes,

$$\Delta_\phi = \Delta_\phi(t) = N_{\phi_1}(t) - N_{\phi_2}(t) \tag{3.6}$$

where $N_{\phi i}(t)$ are the phase noise terms of the lasers. As is well known, the term $\Delta_\phi(t)$ is a noise process with a variance equal to the sum of the variances of $N_{\phi 1}(t)$ and $N_{\phi 2}(t)$.

Things change drastically if the two lasers are not independent. In fact, let us suppose that the lasers show exactly the same time-varying phase noise process N_ϕ (t). In this case, in the phase term Δ_ϕ of (3.4) and (3.5), the time-varying component (i.e., the noise process) goes to zero, and only a phase constant is left. Thus, the PD generates an RF signal ideally free from phase noise. The action of forcing the two lasers to have the same phase noise is referred to as phase locking, as reported in the scheme of Figure 3.2(b).

In the following, we consider the most common and effective techniques for obtaining phase-locked lasers to be used in the photonics-base generation of RF signals.

3.3.1 Generation of phase-locked lasers through RF modulation

The most straightforward method for generating phase-locked lasers, and hence phase-stable RF signals, considers the generation of one of the lasers directly from the other by modulating one laser (usually referred to as the master laser) with an RF oscillator, and considering one modulation side mode as the second laser. As it is evident, with this method the detuning between the two lasers and their reciprocal stability is directly dependent on the frequency and stability of the RF oscillator modulating the master laser.

Nevertheless, this method shows a couple of useful features that make it particularly interesting.

By exploiting the nonlinearities of the optical modulators, it is possible to generate higher-order optical side modes easily, and these modes can be used to generate RF signals at multiples of the RF oscillator frequency. For example, taking advantage of the sinusoidal electro-optical transfer function of the Mach–Zehnder

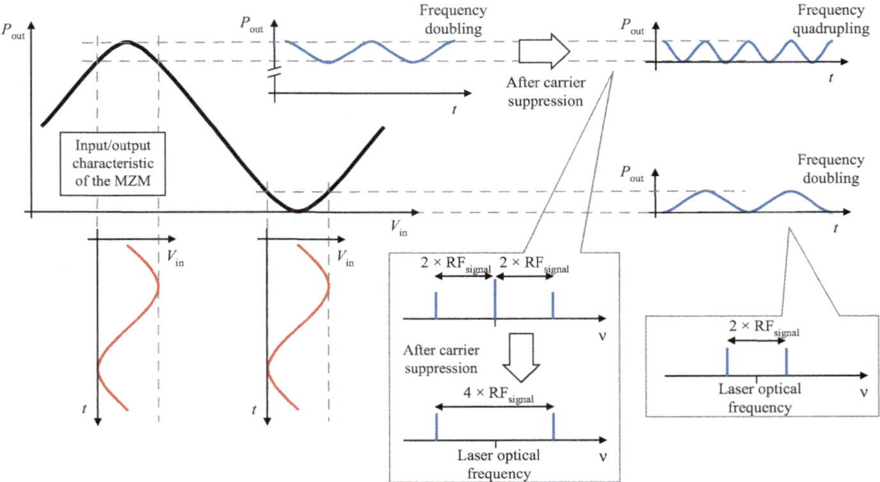

Figure 3.3 Photonics-based frequency doubling and quadrupling, exploiting the nonlinear electro-optical transfer function of the Mach–Zehnder modulator

modulator (MZM), it is straightforward to implement the doubling or quadrupling of the RF modulation frequency by opportunely setting the bias voltage of the modulator to the minimum or to the maximum transmission point (in the case of frequency quadrupling, it is also necessary to suppress the original optical carrier) (Figure 3.3) [10].

With a similar approach, it is also possible to realize optical frequency combs (OFCs), that is, optical spectra composed of several laser lines, all equally spaced and phase-locked with each other. OFCs can then be used either as laser sources (e.g., as lasers for wavelength division multiplexing systems [11]) or as sources of RF signals at any multiple frequency of the RF oscillator [12]. OFCs can be obtained by modulating the master laser in a cascade of intensity and phase modulators. By carefully setting the power of the driving signals, it is possible to obtain a large and flat OFC, as the one reported in Figure 3.4 [13–15].

It is worth noting that, in order to exploit an OFC to generate RF signals at a desired carrier frequency, it is necessary to isolate two laser lines. This operation can be very challenging, in particular if the RF oscillator is at frequency below few gigahertz, so that usual optical filters are not sufficiently selective to pick the desired laser line and suppress the others. Alternatively, the entire OFC can be detected in a PD, thus generating beatings at all the multiples of the comb line spacing. Then, an electrical filter can be used to select the beating at the desired frequency in the RF domain.

As said above, the reciprocal phase stability of the generated comb laser lines depends on the quality of the RF oscillator, and so is for the generated RF signal at

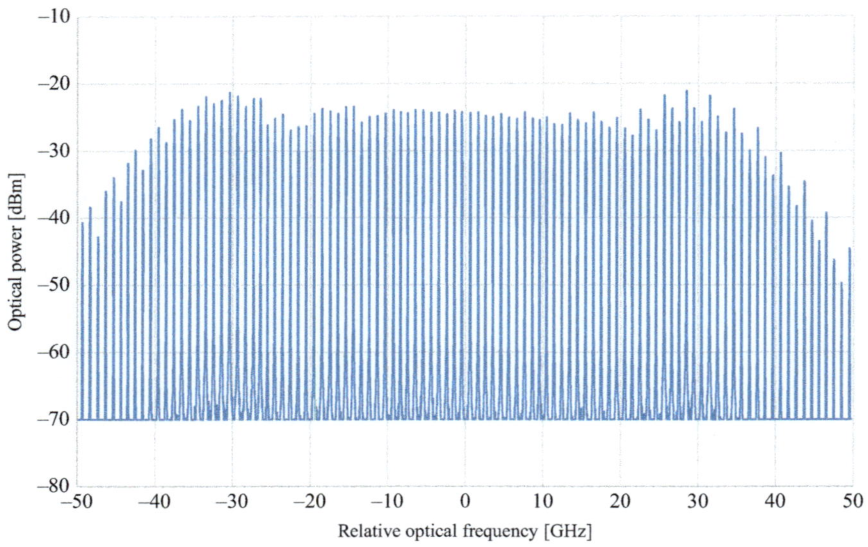

Figure 3.4 Example of optical frequency comb generated as in [15], obtaining more than 80 laser lines in a range of 10 dB, with a frequency spacing of 1 GHz

multiple frequency. It is important to underline that the frequency multiplication also affects the phase noise: the laser line generated as the second harmonic of the oscillator holds twice the oscillator phase noise, the third harmonic three times, and so on. This noise multiplication must be taken into account when generating RF signals this way, and thus the RF oscillator must be chosen with a very low intrinsic phase noise.

3.3.2 Laser phase locking through injection locking

Let us have two lasers, one of which (the master laser) is injected into the cavity of the other (the slave laser). If the wavelength of the CW optical signal from the master laser is sufficiently close to the resonance wavelength of the slave laser, then the slave laser is forced to emit at the same wavelength and at the same phase of the injected laser [16]. This mechanism is called injection locking.

The quality of the locking operation depends on a large number of parameters, mainly the power of the injected light and the detuning between the wavelength of the master laser and the nominal wavelength of the slave laser. The higher the injected power and the closer the two nominal wavelengths, the more precise the locking operation. It is also useful to introduce the parameter defined as locking range of the slave laser, that is, the maximum detuning between the slave and the master that allows the slave to lock. This parameter also depends on the injected power: the higher the injected power, the larger the locking range. Another important parameter affecting the quality of the locking mechanism is the linewidth of the two lasers. The narrower the linewidth, the better the locking, but also the narrower the locking range.

In order to generate an RF signal exploiting the injection locking, we need to have two detuned lasers, and a way for deriving a master laser from one of those, so that its wavelength is the same as the nominal one of the slave and it can be used for injection. A possible configuration is considering a laser with multiple phase-locked modes as, for example, a Fabry–Perot (FP) laser. So, by injecting one of the modes into the slave laser, the latter will be locked not only to the injecting mode, but also to any other longitudinal mode of the multimode source.

The quality of the obtained RF signal depends on the quality of the locking mechanism, hence on the injected power, on the master–slave detuning, and on the linewidth of both the multimode laser and the injected laser [17]. Unfortunately, multimode lasers are rarely low noise and they usually present a large linewidth. Therefore, the RF signals generated using the injection locking technique are usually only fairly stable in phase. Moreover, the generated RF frequency, which is determined by the detuning between the multimode laser and the slave, depends on the mode spacing of the master.

It is important to note here that the injection locking technique can be seen as an active optical filtering, selecting and amplifying the injected tone (the one at the correct wavelength) out of several ones. Therefore, it can be used also in conjunction with the OFCs described above, to select laser lines out of a very large comb.

Figure 3.5 reports the phase noise curve of an RF signal at 3 GHz generated by heterodyning a master laser with the third line of a large comb obtained by

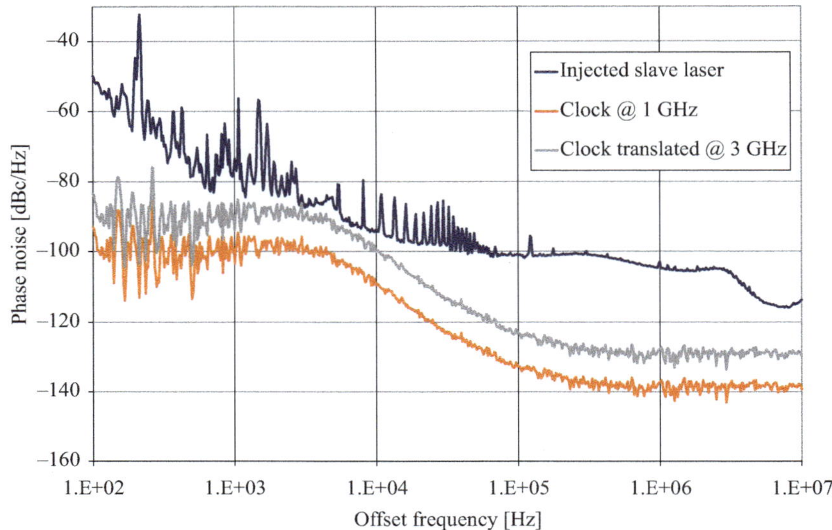

Figure 3.5 Example of phase noise obtainable by exploiting the OFC generation and the injection locking for the selection of the comb modes

modulating the master in a cascade of optical modulators driven by an RF clock at 1 GHz, as in Figure 3.4 [15]. The third line is extracted from the comb by injection locking a slave laser (here, it was the same kind of laser used for the master, with a linewidth of 100 kHz). The phase noise of the RF signal thus generated is compared with that of the original clock signal at 1 GHz. As can be seen, the phase noise of the signal at 3 GHz is significantly higher than that of the RF clock at 1 GHz. Part of this difference comes from the different frequency between the photonics-generated signal and the clock. Since the phase noise depends on the square of its nominal frequency, the phase noise curve of the clock at 1 GHz must be translated 9.5 dB up to be comparable with the 3 GHz signal. As can be seen, even after translation, the noise difference is still significant.

3.3.3 Opto-electronic oscillators

RF signals with excellent stability can be achieved using opto-electronic oscillators (OEOs). These are laser cavities modulated by a self-sustained opto-electronic feedback that allows obtaining very pure oscillations [18], capable of satisfying the most demanding applications.

Since in the OEO the phase noise rejection is directly proportional to the squared delay length of the opto-electronic feedback [19], the first proposed OEOs [20–23] included a long fiber delay. Their basic architecture is shown in Figure 3.6(a). It consists of a CW laser that is fed into an intensity modulator. Its output is then passed through a long optical fiber (in the order of $>$ 1 km) and detected by a PD. Then, the recovered electrical signal is amplified and filtered out by a band-pass filter, and sent back to the modulator input in order to complete the optoelectronic cavity. When the gain of the cavity is greater than the losses, the OEO begins to oscillate at the frequency selected by the filter.

Since a BPF determines the oscillating frequency in this kind of OEO, a frequency tuning is not possible. To enable it, the electrical filter must be substituted by an optical filter, which has much higher tuning capabilities. One of the most promising optical tunable filter for OEOs is based on whispering optical cavities, which are highly selective, with quality factor in the order of 1 million and above, and are extremely compact [24–27].

There are many factors contributing to OEO phase noise. Environmental fluctuations affects the acoustic range (1–10 Hz), while from 10 Hz to 1 kHz the flicker noise of the RF amplifier is the main contribution, and above 1 kHz white noise and laser noise are dominating. A significant improvement can come from removing the microwave amplifier [28]. This amplifierless OEO requires very high EO link gain, which means using a high-power laser coupled with a high-power-handling PD, and highly efficient modulators. Optical amplification can also be introduced to this purpose.

As said, in an OEO the longer the OE loop, the lower the phase noise. However, increasing the loop length (or the filter quality, in the case of circulating architectures) lowers the frequency of the cavity modes, and unwanted cavity modes can pass through the loop filter as a spurious tone. To suppress this problem, a second

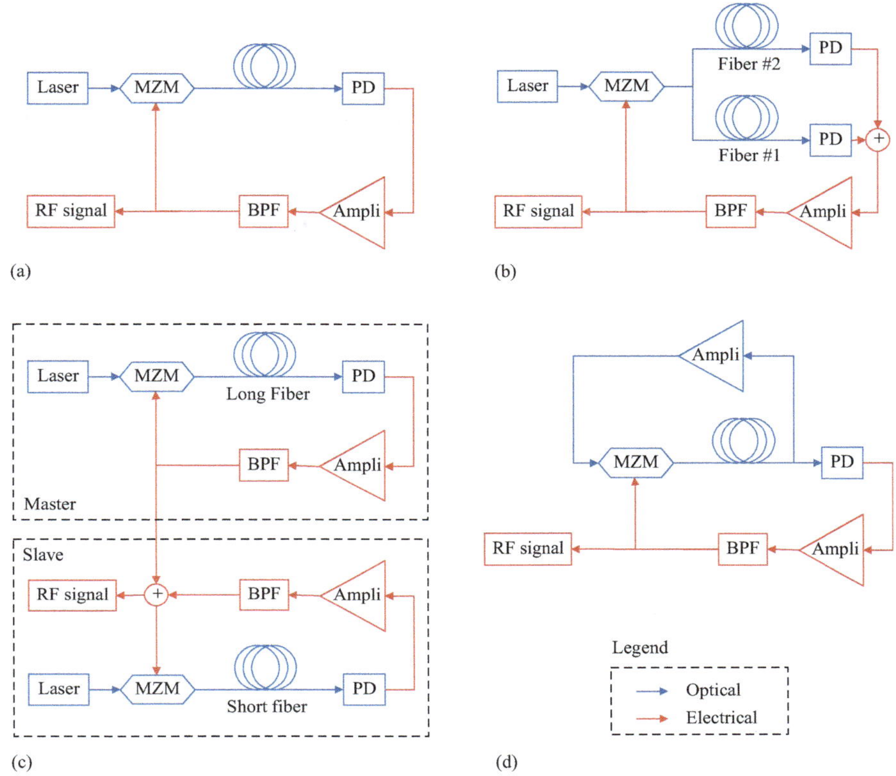

Figure 3.6 *(a) Basic scheme of an OEO including a long fiber loop; (b) OEO with a second OE loop to suppress the cavity modes; (c) OEOs in master/slave configuration; (d) coupled OEO, generating ultra-stable optical pulses*

loop is introduced (Figure 3.6(b)) [29–32] so that the modes of the complex cavity become the minimum common multiple of the modes of the two loops.

A similar approach is implemented in the master–slave configuration [33,34] (Figure 3.6(c)), where a short loop OEO (slave) is injection locked by a longer loop OEO (master). Finally, if the CW laser pump is replaced with a recirculating optical amplifier, such as an Erbium-doper fiber amplifier (EDFA) or a semiconductor optical amplifier (Figure 3.6(d)), a coupled OEO (COEO) is realized, that produces optical pulses [35–38]. As in the dual loop and in master–slave configurations, the circuit is composed by two loops, being one all-optical and the other opto-electronic.

In terms of performance, the basic OEO and the COEO have demonstrated phase noise in the order of –115 dBc/Hz at 10 kHz offset for a generated RF signal at 10 GHz, while the master–slave OEO configuration has reported a –150 dBc/Hz at 10 kHz for the same RF frequency.

3.3.4 Mode locked lasers

The laser schemes generating optical pulses through an all-optical feedback, as the last example described in the paragraph above, are also referred to as mode locked lasers (MLLs) [39]. In fact, these schemes generate pulses by forcing the modes of an optical cavity to have exactly the same phase (mode locking), so that the optical signal can be described as a sum of several cavity modes, all in-phase to each other, which gives as an output an optical pulse train at a repetition rate equal to the mode spacing. The optical spectrum of these lasers presents a large number of equally spaced modes, all with the same phase, and the spacing $\Delta\nu$ depends on the cavity length:

$$\Delta\nu = c_n / L_{\mathrm{RT}} \tag{3.7}$$

with c_n the speed of light in the medium, and L_{RT} the round-trip length of the cavity (in the case of a ring laser, L is the length of the ring; in the case of a linear cavity, L is twice the length of the cavity).

The MLLs have been relegated to the laboratories of physical departments for a long time. In the early year 2000, they became popular within the optical communication research for optical time-domain multiplexing. More recently, they have found an important niche of use for absolute frequency measurements, in association to the multi-octave supercontinuum generation [40].

The basic scheme of an MLL considering a ring cavity is reported in Figure 3.7(a). The scheme considers a gain medium (which could be either a semiconductor amplifier or a fiber-based amplifier) and a saturable absorber (SA), and is usually referred to as a passive MLL. At steady state, a single optical pulse is circulating in the ring, and the SA lets the peak of the pulse passing with low attenuation, while the tails of the pulse encounter a higher loss. Therefore, the SA forces the generation and maintenance of the pulse. In the spectrum-based description, the SA modulates the attenuation of the cavity, forcing the modes to assume the same phase (i.e., all the modes are modulated by the SA in order to see a reduced cavity loss at the same

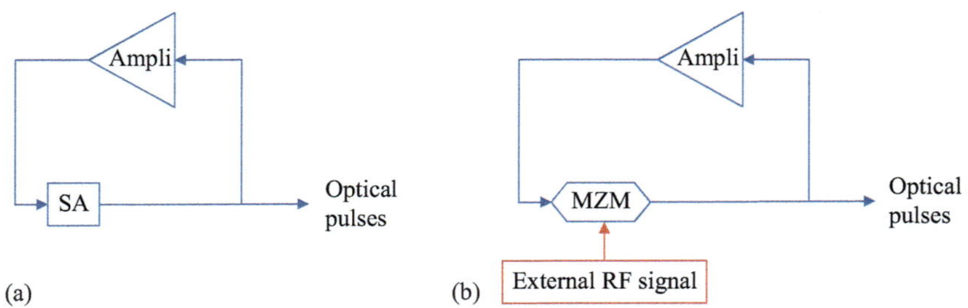

Figure 3.7 (a) Basic scheme of a passive mode-locked laser; (b) scheme of an active harmonic MLL (the frequency of the external RF signal must be an integer multiple of the cavity fundamental mode)

instant). Then, an optical coupler extracts a portion of the optical power to make it available.

MLLs can easily generate pulses with a duration below 1 ps. Their spectrum can therefore exceed several nanometers. In order to minimize the pulse width and maximize the bandwidth, a fine control of the cavity chromatic dispersion is needed to avoid propagation delays from mode to mode. In some implementation, non-linear effects as four-wave mixing and cross-phase modulation are also exploited to further reduce the pulse duration [41].

Starting from the base configuration described above, several architectural modification can be implemented. For example, a piezo-activated tunable delay can be inserted in the cavity to lock the MLL repetition rate to an external clock, and an optical filter can be inserted to force the MLL to oscillate in a specific spectral region. The laser cavity can also be made oscillating at a frequency higher than the fundamental one, so that the repetition rate is higher, and the spectral mode spacing is larger. This can be done by substituting the SA with an electro-optic modulator (e.g., an MZM) and driving the modulator with an external RF signal at a multiple of the fundamental frequency of the cavity, implementing a so-called active harmonic MLL (Figure 3.7(b)). The RF signal driving the modulator can also be derived directly by the cavity itself, implementing an electro-optic feedback including a PD, an RF filter to select the desired harmonic of the cavity, an amplification stage, and a phase controller to correctly drive the modulator. This scheme corresponds to the description of the coupled OEO of the previous paragraph (Figure 3.6(d)), and is also called regenerative MLL. As for the OEOs, the longer the cavity the lower the phase noise, but then the problem of the spurious cavity modes arises, which can be treated as discussed above for OEOs.

MLLs have demonstrated very high performance in terms of stability, comparable with that obtained by OEOs, therefore they have been largely proposed for the generation of RF signals [42–46]. A significant difference between the two architectures comes from the fact that OEOs generate a single RF frequency, while MLLs can generate virtually any RF signal with a frequency multiple of the repetition frequency [42–46].

It is fundamental to underline here that all the RF signals generated by the same MLL are coherent with each other, that is, they have exactly the same phase, and their phase noise shows the same behavior. Clearly, since all the possibly generated RF signals are multiples of the RF signal at the MLL repetition rate, the phase noise is multiplied as well, as can be seen in (3.8), where we consider $\mathrm{RF}_n = n\cdot\mathrm{RF}$:

$$S_{\mathrm{RF},n}(t) = A_{\mathrm{RF},n}(t)\cos\left[2\pi n\nu_{\mathrm{RF}}t + nN_{\phi_{\mathrm{RF}}}(t)\right] \tag{3.8}$$

This is evident from the measure reported in Figure 3.8, where RF signals at 10, 20, 30, 40, and 50 GHz have been generated from a regenerative MLL at 10 GHz. Their phase noise shows the same shape, but the curves appear translated 6 dB for every frequency doubling [47].

It is important to underline that the measured phase noise at 10 kHz offset for the signal at 10 GHz is –125 dBc/Hz.

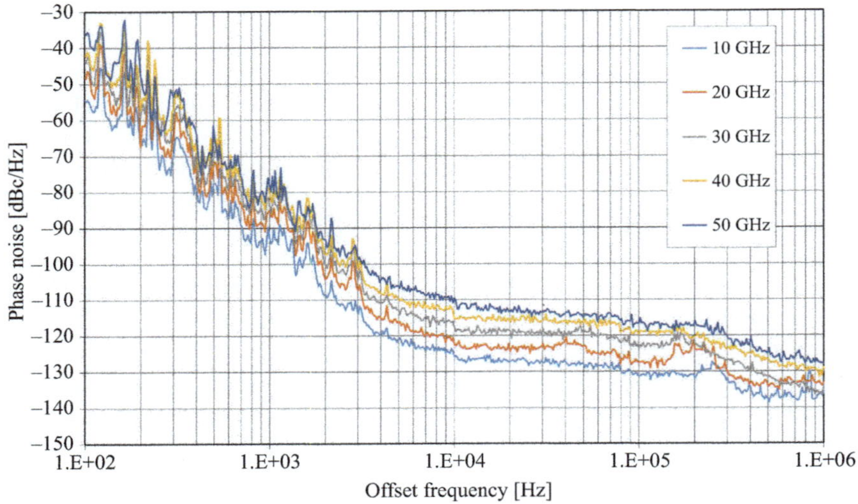

Figure 3.8 Measured phase noise of RF signals at 10, 20, 30, 40, and 50 GHz generated by a regenerative fiber-based MLL at 10 GHz repetition rate

Thanks to the extremely large number of available phase-locked modes from MLL, they have been also proposed for simultaneously generating modulated wideband RF signals with agile carrier frequencies selection [48,49].

3.4 Photonics-based RF detection

The detection of RF signals in a software-defined environment would require the availability of ADCs with an input bandwidth extending from DC to the MMW, with a consequently fast sampling rate. Recently, pushed by the requests of the communications sector, similar ADCs are hitting the market, thanks to the advances in electronics and to the time-interleaved sampling technique. For example, commercial real-time oscilloscopes are now available with an input bandwidth up to 100 GHz and a sampling rate as high as 240 GSps. Unfortunately, the extremely large noise bandwidth and the architectural complexity required for reaching such high sampling rates limit the maximum signal-to-noise ratio (SNR) of the detected signals to about 30 dB, and this value is too low for other fields of use as the radar or EW systems, where SNRs higher than 40 dB are needed. For these applications, the detection of RF signals is still managed exploiting RF down-conversion and reliable ADCs at slower sampling rates.

Figure 3.9(a) reports the conventional scheme for an RF down-conversion. As for the RF up-conversion structure, it exploits multiple stages composed of a mixer driven by an LO and followed by a BPF in order to move the detected signals into the bandwidth of the ADCs. As in the case of the transmitter, each of these stages

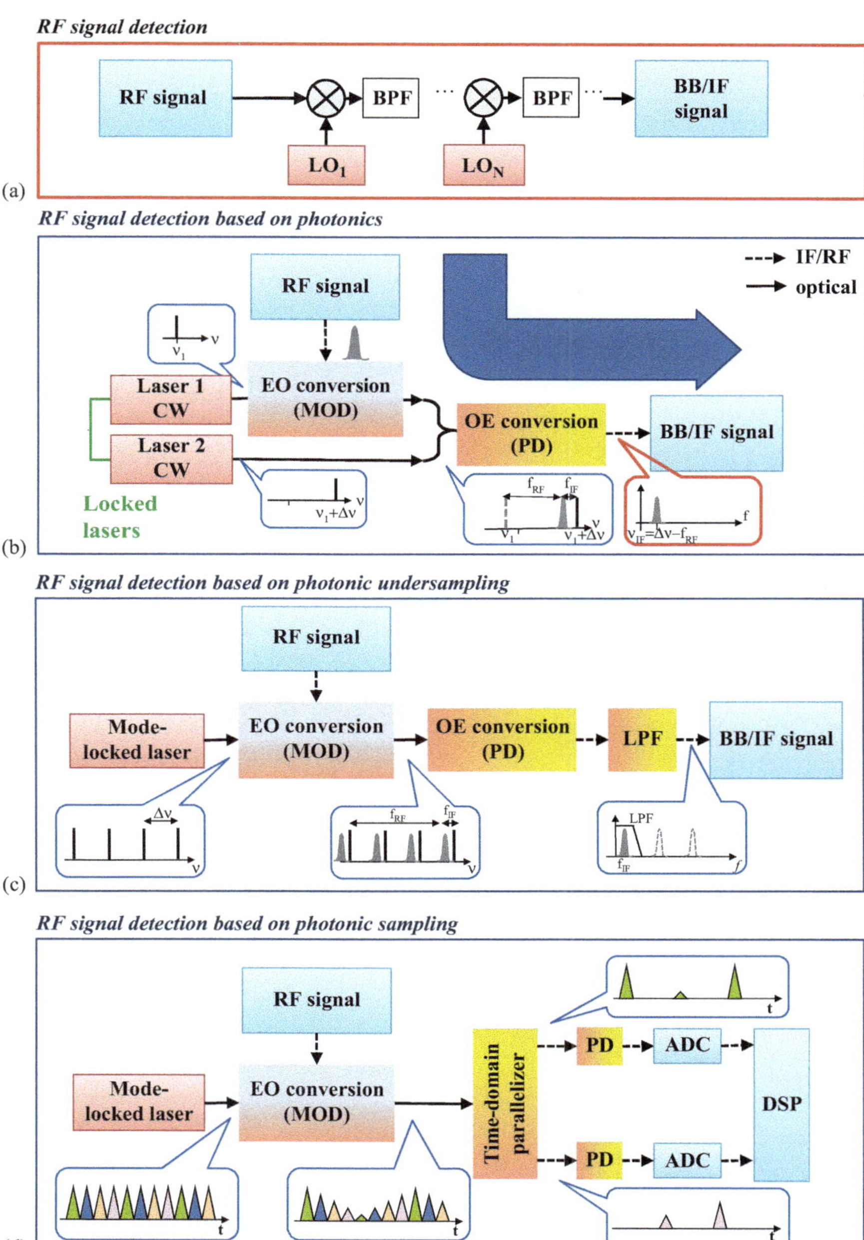

Figure 3.9 (a) Conventional RF down-conversion scheme; (b) RF detection based on photonics down-conversion; (c) RF detection based on photonic undersampling; (d) RF detection based on photonic sampling and time-demultiplexing

introduces phase and amplitude noise, and this noise increases with the detected frequency. Moreover, this architecture is very frequency-specific and does not allow easily changing the input RF frequency.

The photonics-based RF detection exploits the huge electro-optical bandwidth of optical modulators to translate the RF signal to the optical domain and then back to the electrical domain, so that its acquisition and digitization can be realized with more flexibility and precision [6–10,17,18,42–46,48–52]. Several different schemes for RF receivers have been proposed so far. In the following, we review the main techniques, highlighting their peculiar strengths.

3.4.1 RF detection by photonic down-conversion

Analogously to the RF generation through direct up-conversion, the detection of RF signal can be implemented through direct down-conversion. This can be effectively realized exploiting again the concept of heterodyning [53], in conjunction with the generation of phase-locked laser lines, as analyzed in the previous paragraph (Figure 3.9(b)).

It is of particular interest to consider the direct down-conversion when using a comb of phase-locked lasers as the spectrum of an MLL. In the scheme reported in Figure 3.9(c), every comb laser is modulated by the RF signal, generating sidebands at f_{RF} from the carrier, which are detuned f_{IF} from the closest comb line. Therefore, detecting the entire spectrum by the PD, all the modulation sidebands beat with its closest laser line, generating a contribution at f_{IF}. All these contributions sum up in phase at the output of the PD, giving the direct down-conversion of the detected RF signal to IF. Finally, after a low-pass filter, the down-converted signal can be digitized by an electrical ADC [54]. This operation is therefore a form of undersampling, translated in the photonics domain. With this approach, RF signals with carrier frequencies up to 40 GHz and bandwidth of 20 MHz have been digitized in ADC at 100 MSps with an effective number of bit equal to 7 [55].

The scheme based on the MLL is also capable of down-converting multiple RF signals at different frequencies simultaneously. In fact, if we consider to modulate every laser line with two RF signals at different frequencies, two orders of sidebands are generated, respectively, at f_{IF1} and f_{IF2} from the closest laser line. So, as in the case of detecting a single RF signal, when the modulated MLL spectrum is received by a PD, two orders of components are generated, which are directly down-converted at IF1 and IF2. This is feasible provided that the corresponding down-converted IF replicas do not overlap [54].

3.4.2 RF detection by optical sampling

The optical pulses from an MLL can also be used as a high-rate sampling system to directly sample broadband and high-frequency RF signals [50,52,56]. With respect to the direct RF sampling, this technique shows a larger RF input bandwidth and can reach much lower sampling jitter (which is the limiting factor in electronic ADCs). As an example, Figure 3.9(d) shows the sampling scheme used in [54,56]: the optical sampling occurs when the RF signal modulates the MLL pulses; from that moment, the amplitude of the modulated pulses represents the optical samples (the sampling rate and its jitter are given by the MLL). In order to digitize the samples by means of precise ADCs, the optical samples are then demultiplexed in the time domain in a fast optical switching matrix in order to parallelize the samples to lower sample trains that are synchronously digitized by precise electronic ADCs. The acquired samples must then be interleaved in the digital domain. With this technique, an SNR higher than 42 dB has been demonstrated for signals at 40 GHz [56].

3.4.3 Other photonics-based RF receiving techniques

Several other techniques have been proposed to acquire RF signals using photonics. Among those, the "time-lens" approach is worth mentioning. This technique solves the problem of having ADCs with simultaneously sufficient sampling rate and precision, by stretching the optical signal in the time domain through a combination of dispersive fibers, so that the output optical signal is slowed down and can be sampled in an easier way (i.e., with low-sampling rate precise ADCs).

Due to the time dilatation, it is difficult to exploit this method with arbitrarily long RF signals. On the other hand, radar systems usually work with short RF pulses, and therefore they could take advantage of the "time-lens" approach. Moreover, it is worth mentioning the work reported in [57], where an impressively wide RF band from DC to 48 GHz has been continuously acquired (i.e., as without time-dilatation) thanks to a wavelength-division multiplexing approach, with a total sampling rate of 150 GSample/s.

3.5 Photonics RF signal transport and distribution

Once the RF signal is loaded on an optical carrier, it becomes convenient to use optical fibers to transport the signal, implementing a radio-over-fiber (RoF) system. In fact, fiber transmission is broadband, low-loss (as low as 0.2 dB/km, while RF waveguides have a propagation loss in the order of several dB/m), and EMI free. Therefore, the optical fiber allows transporting the RF signal over long distances without significant distortions. This can be fundamental, for example, for remotizing the antennas in a wireless or radar system. Moreover, optical fibers are light, small, and flexible, and therefore can be installed in complex and narrow sites as, for example, in unmanned military vehicles and in satellites.

The performance metrics of a RoF system are the same as for standard RF links and are summarized by a set of relevant figures as gain, noise figure (NF), spurious free dynamic range (SFDR), third-order intercept, to name the most relevant.

The basic RoF scheme is implemented by modulating the RF signal on the amplitude of a laser, sending it through optical fiber, and detecting it directly with a PD (intensity modulation, direct detection) at the other side of the fiber link. The amplitude modulation can be realized either by direct modulation of the laser

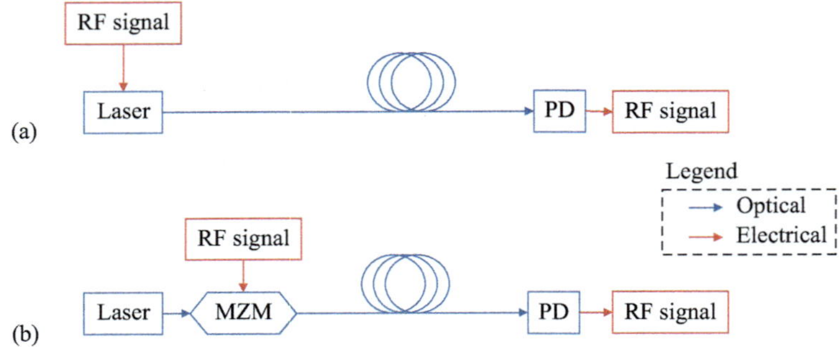

Figure 3.10 Scheme of a RoF system based on (a) direct modulation; (b) external modulation

source (Figure 3.10(a)) or by external modulation through, for example, an MZM (Figure 3.10(b)).

Other approaches can be exploited (e.g., modulating the analogue signal on optical phase or polarization, or implementing a coherent detection by beating the modulated carrier with a second optical carrier, rather than simply squaring it in the photodetector). By now, here we focus on the amplitude modulation only as this is the widely deployed scheme.

Considering links with neither electrical nor optical amplification, the gain range reported in the literature is in the order of <–20 dB and largely spreading due to the different components, modulation techniques, and specific architectural solutions. Gain in directly modulated links is limited by the laser modulation efficiency (which ranges around 0.1–0.3 W/A). In externally modulated links, instead, gain can be more easily increased acting mainly on the laser power, and on the modulator's on–off switching voltage (usually referred to as $V\pi$). By exploiting high-power lasers and low $V\pi$ modulators, modulation efficiency higher than 10 W/A and positive link gain were reported [58].

The NF strongly depends on the link gain and on the intrinsic noise characteristics of the components (e.g., the laser relative intensity noise [RIN]). If a low noise amplifier is introduced in the scheme before the link, a significant improvement in NF and gain can of course be achieved. SNR is affected by the noise sources in the links, mainly the RIN of the laser, the shot noise of the PD, and the thermal noise. At high photocurrents (i.e., at high optical power), the SNR turns out to be limited by the laser RIN performance. Finally, the SFDR provides an overall picture of dynamic range and noise characteristics. SFDR is typically limited by the nonlinearities of the modulator and of the photodetector, as well as by the link output noise. The best reported values (around 134 $\text{dB}\cdot\text{Hz}^{2/3}$) were achieved by using linearized modulators [58].

Commercial RoF systems currently ensure an analog bandwidth up to 20 GHz with a gain of –25 dB (external modulation, no amplification), an SNR up to 146 dB/Hz, and an SFDR up to 109 $\text{dB}\cdot\text{Hz}^{2/3}$.

Considering the RoF system applied to a radio system as, for example, a radar, the electro-optic and opto-electronic signal conversions can be either managed through the up- and down-conversion schemes described above, or realized by straightforwardly modulating the RF signal on an optical carrier at the optical transmitter side, and then detecting the optical signal in a wideband PD at the optical receiver side. If the fiber link is intended to remote a receiving antenna, as in the case of a radar system, it can be convenient to keep the laser at a base station, send the unmodulated optical carrier to the antenna through an optical fiber, and move to the antenna only an optical modulator, where the carrier is loaded with the received RF signal (Figure 3.11) [59].

In the specific applications to radar systems, a couple of possible issues arises with RoF solutions. The first one is related to the available linear dynamic range of the transport system. In fact, applications as radars or EW systems require a very broad dynamic range, while the RoF systems using external optical amplitude modulation are often subject to nonlinearities induced by the modulator itself. In order to increase the maximum linear dynamic range, RoF systems exploiting phase modulation (PM) have been developed. While this kind of modulation is very linear, the opto-electronic conversion of the RF signal becomes more complex, requiring either a frequency discriminator for implementing a direct detection or an even more complex coherent detection scheme [59].

A second issue is the effect of chromatic dispersion when the RF signal is transferred on the optical carrier by means of a double sideband amplitude modulation. In fact, if the total chromatic dispersion of the fiber link is high enough (e.g., due to a long fiber link), some frequency components of the two sidebands can undergo a significantly different phase shift due to the dispersion, which gives a notch in the transmissivity of the system at those frequencies that are turned exactly out-of-phase. In order to suppress the fading, it is possible either to make use of PMs or to exploit the single sideband (SSB) amplitude modulation. Once the effect of the fading is suppressed, RoF systems can be used to cover spans as long as several hundred kilometers, also thanks to the use of optical amplifiers (EDFAs). Moreover, RF signals can be distributed across networks, for example, for the precise clock distribution in complex systems as synchrotrons, but also for the distribution of the RF signals in radar networks [60].

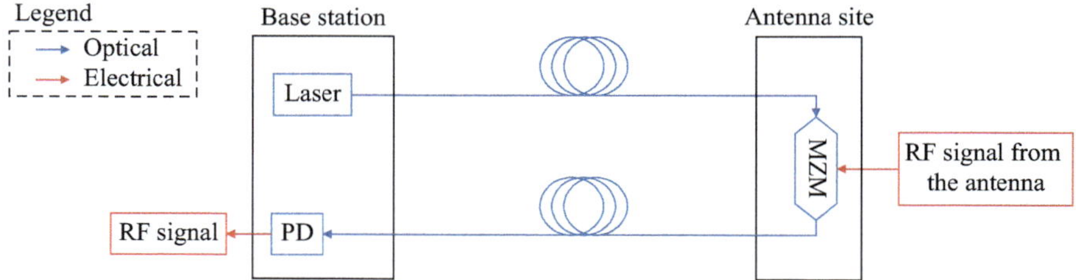

Figure 3.11 Scheme of a RoF for remoting a receiver antenna

3.6 Optical filtering for microwave signals

Filtering is the most common “processing” operation for RF signals. While electronic filtering is a largely established function for low-frequency signals, when the signal frequency increases it becomes a complex issue. In fact, filters parameters such as the flatness, the edge steepness, and the out-of-band rejection worsen with the central frequency, and complex high-performing cavity filters become necessary. These are usually fixed filters, and thus they reduce the flexibility of the overall system. Considering radar applications, this translates in a strong limitation to the software-defined approach. It also has heavy consequences in EW systems, as explained in more details in Chapters 2 and 7.

On the other hand, tuning a filter in the photonic domain is much easier. This becomes clear when considering that an optical filter selects an electromagnetic field (i.e., the optical signal) with a carrier frequency close to 200 THz, and that the maximum required filter tuning range for microwave photonics applications is of few tens of gigahertz: the required tunability is only a tiny fraction (in the order of 10^{-4}) of the filter central frequency!

The generic scheme of a microwave photonic filter is sketched in Figure 3.12. The RF signal is first transferred to the optical domain by modulating a laser. Let us assume that a single-sideband modulation is implemented. The spectrum of the modulated optical signal is composed of the carrier and the modulation sideband, which is a copy of the RF signal translated to optical frequencies. An optical tunable filter then is used to select a portion of the modulation sideband. After reintroducing the optical carrier and heterodyning, the PD gives back the RF signal filtered by the optical filter.

The optical filter can be easily tuned changing some of the physical characteristics in the filter structure. This can be done, for example, through thermal effects (with tuning times in the order of few hundred microseconds or more), or through the much faster carrier injection (with tuning time down to one nanosecond [61]). Another tuning possibility is changing the wavelength of the laser, while keeping fixed the filter position.

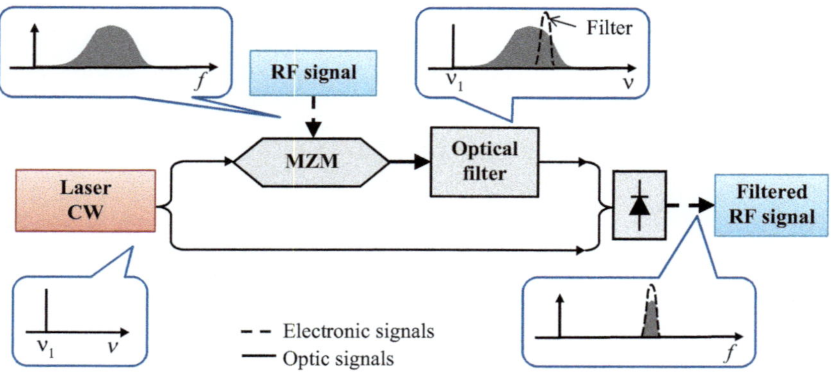

Figure 3.12 Basic scheme of a microwave photonic filter

The equivalent shape of the microwave filter follows the shape of the optical filter. Therefore, the optical filter must fulfill the filtering requirement of the RF application. Besides the tunability, which is a simple feature to achieve with photonics, several other requirements can be very challenging for an optical filter, again considering that its central frequency is in the order of 200 THz. For example, a filter bandwidth of 1 GHz corresponds to a quality factor of about 200,000!

To reach the technical specifications required for RF systems, microwave photonic filters are commonly implemented by interferometric structures [62]. A typical example is the FP filter, which is realized by placing two semi-transparent mirrors in front of each other: the light entering the structure tends to be trapped in and accumulates, only if the wavelength of the light is an integer multiple of the optical path length inside the resonator. In fact, under this hypothesis, every new bunch of light entering the structure sums up in phase with the trapped light. On the contrary, wavelengths that do not respect this hypothesis instead add incoherently and tend to fade. At the output of the filter, a fraction of the accumulated light is transmitted. Therefore, the filter in/out characteristic shows a transmission peak at the resonant wavelength, while the other wavelengths are rapidly attenuated.

The 3-dB bandwidth of the transmission window is an important parameter of the FP filter. It depends on the quality of the resonator, that is, how long the light can circulate in the cavity. The longer the photon lifetime in the cavity, the narrower the bandwidth. Therefore, the 3-dB bandwidth of the FP filter is affected by the reflectivity of the semi-transparent mirrors, and by the intrinsic losses of the resonating cavity.

Several other types of optical filters can be described by comparison with FP filters. For example, fiber Bragg gratings (FBGs) can be seen as FP filters where the semi-transparent mirrors are substituted by Bragg gratings. FBGs offer the possibility to realize very long Bragg gratings, that is, very reflective mirrors, so that the filter bandwidth can be smaller than for FP filters.

Another similar architecture is the microring resonator (MRR) (Figure 3.13(a)). This structure is made possible by the advances of integrated photonics and consists in a waveguide in ring configuration that is weakly coupled to a feeding waveguide. The light at the resonant wavelength is trapped in the microring and is not released,

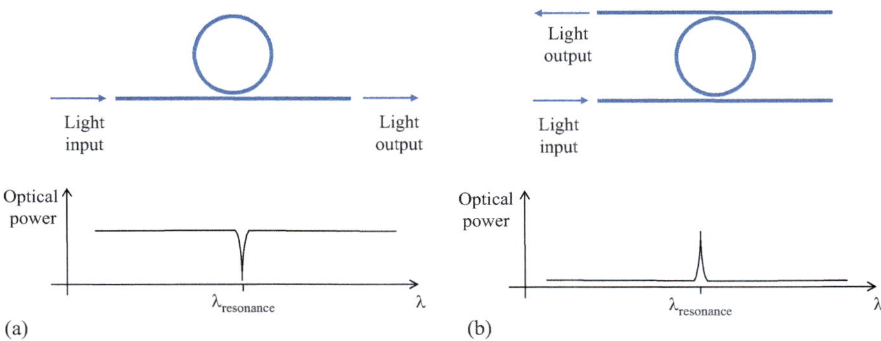

Figure 3.13 Scheme of an MRR: (a) in all-pass filter configuration; (b) in add/drop configuration

realizing a notch filter. This resembles an FP filter with one of the mirrors characterized by having a total reflectivity. Given the structure of the filter, this is referred to as a MRR in all-pass filter (APF) configuration. If a second waveguide is added, then an add/drop configuration is realized, and the filter can work as a BPF.

By designing the integrated MRR accurately, it is possible to increase significantly the photon lifetime with respect to the FP filter. In particular, the reduction of the intrinsic losses is the most important field of development. This can be done by optimizing the design of the ring waveguide (roughness, radius, index contrast) or by choosing the material with the lowest propagation losses [63]. Examples of the latter approach can be found, where the MRR is substituted with a micro-torus realized in ultra-low loss glass, and the light circulates in the form of whispering gallery modes [64]. This way, impressive quality factors above 1,000,000 have been reached. This result is particularly useful in application as the OEO (see Section 3.2).

The examples above all realize optical filters with a Lorentzian shape, which shows non-flat top and slowly decaying edges. This shape therefore limits the applicability in the microwave fields, where a high selectivity (i.e., the ability of strongly rejecting adjacent frequency components) is crucial. Instead, optical filters with narrow flat-top bandpass, abrupt slope in the transition band, and high stop-band rejection are necessary. To this scope, high-order MRRs-based architectures have been proposed, where multiple MRRs are cascaded in series [50,51], demonstrating bandwidth in the order of 70 MHz, with >50 dB of stop band rejection.

Another possible structure implementing a filter function useful for microwave application is the MRR-loaded Mach–Zehnder interferometer [65,66], which can implement Chebyshev filter shapes of various orders. In the example sketched in Figure 3.14, we report the structure of a Mach–Zehnder loaded with one MRR only, which gives the filtering function of a Chebyshev of order three.

Another possibility is given by the quarter-wave phase-shifted Bragg gratings. In these structures, the perturbation in the center of the Bragg grating is shifted by a quarter of the Bragg wavelength, so that two half-wave FP resonant cavities are created. This improves the single-cavity FP filter by realizing a more rectangular spectral response [67]. Extending this approach with several phase-shift sections, a sharp transmission window within the grating stopband can be tailored, implementing a box-like transfer function. An example of PSBG filter of the 10th-order (i.e., comprising 10 phase-shift sections) with 650 MHz bandwidth and rectangular transmission shape has been reported based on FBGs [68]. On the other hand, interesting results are being obtained with integrated photonics, in particular allowing an ultra-compact device with six phase-shift sections and an out-of-band rejection >35 dB [69] (Figure 3.15).

The optical filters described above can all implement a tunable microwave filter if the associated laser can be tuned (see Figure 3.12). Anyway, some of the described implementations can be tuned themselves. For example, the structures using MRRs can be tuned either thermally or by carrier injection (provided that they are realized with doped, conductive semiconductor materials). This is fundamental because the fast and wide tuning range of photonics-based filters is an

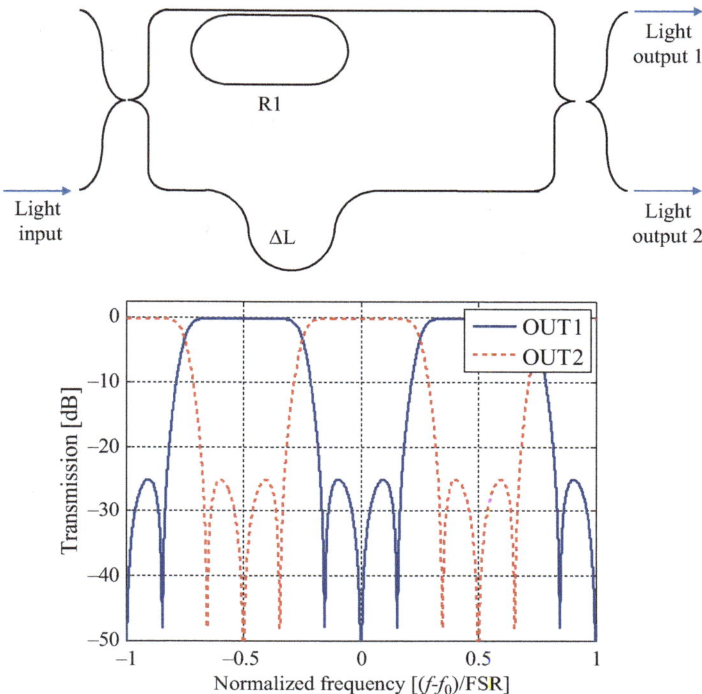

Figure 3.14 Scheme of an MRR-based MZ interferometer and an example of filtering function

essential feature in EW systems to perform adaptive scanning of the RF spectrum in all the applications were situational awareness is critical.

Moreover, it is very important to underline that the filter structures comprising cascades of MRR can be reconfigured, implementing software-defined microwave filters that can tune in frequency and also in shape [70]: this is a unique feature that photonics can bring to the microwave field and could open up completely novel and unexplored possibilities.

3.7 Optical beamforming of RF signals

The beam forming of RF signals in PAAs (also called active electronically steered antennas) allows steering the transmitted RF beam without physically moving the antenna. This solution is used in an increasing number of applications including radars, EW, and communications since it permits a strong reduction of size and weight in the antennas (no moving parts are needed!). In PAAs, the time of emission of the signal at each antenna element is controlled so that the wavefront of the signal generated by the entire antenna array is synthesized to propagate in the desired direction.

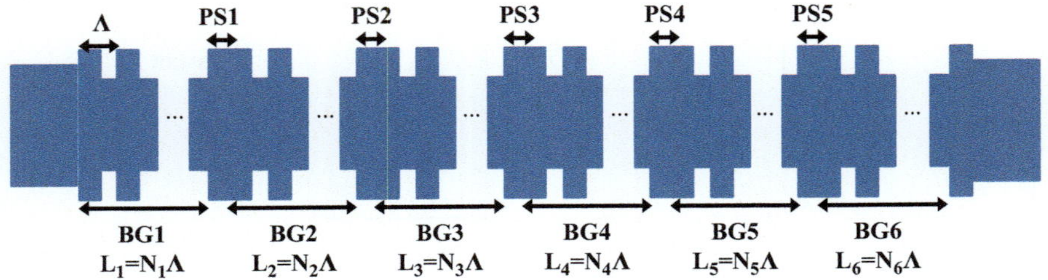

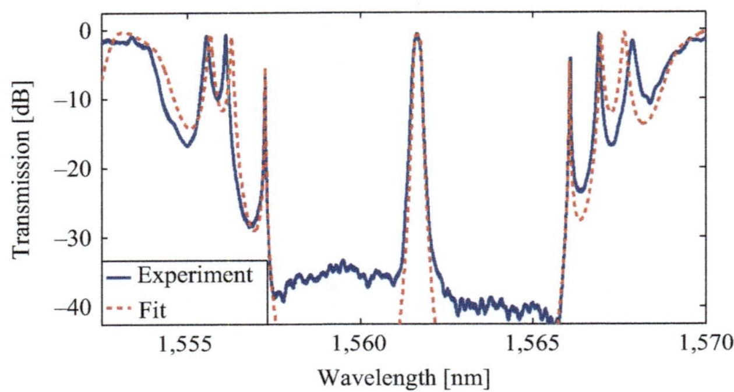

Figure 3.15 Phase-shifted Bragg grating: structure and filter response

The time of emission of the signal at each antenna element is commonly controlled by means of electronic phase shifters since a phase shifting is a correct approximation of a time delay if the delay is small and the signal bandwidth is much narrower than the carrier frequency. In fact, under these hypotheses one can describe a time delay in terms of fractions of the carrier period. When the PAA is transmitting broadband signals, the approximation above does not hold anymore: in fact, in this case a constant phase shifting of all the spectral components in the broad signal spectrum would give different frequencies a different delay. Therefore, when the signal is broadband, the phase shifting approach induces a beam squinting [2]: different frequencies of the signal spectrum aim at a different angle, losing directivity and gain in the antenna. In order to avoid it, a true time delay (TTD) must be controlled at each antenna element.

In current electronically controlled PAAs, the phase shifting is realized by means of analog RF phase shifters. In case of broadband signals, instead, the TTD is implemented in the digital domain (rather than in the analog RF domain) by processing the numerical signal at each antenna element in order to synthesize a delay on the samples. This operation requires a huge digital processing capability and is usually reserved for high-performance applications only.

If the RF signal is transferred in the photonic domain, it is easy to implement either a phase shifting or a TTD, taking advantage of the huge bandwidth, frequency flexibility, and EMI insensitivity of photonics. For both the phase shifting and the TTD, several different solutions are available. In the following discussion, we will describe these methods considering the beamforming in transmission, but the same approaches can be used for controlling the direction of detection in receiving PAAs.

Let us consider an RF signal loaded on a laser by means of an SSB modulation. To implement a phase shifting to the optically carried RF signal, it is necessary to shift the optical carrier with respect to the sideband. This way, once the optical signal is converted back to the RF domain by a PD, the variation in the phase difference between carrier and sideband is transferred to the RF signal. This can be realized by separating carrier and sideband in an optical de-interleaver (as the MRR-loaded MZ interferometer presented above), and shifting the carrier alone, for example, in a phase modulator, before recombining the carrier and the sideband (Figure 3.16(a)) [71,72]. Another possible way for shifting the carrier only is to use a wavelength-specific phase shifter. For example, an MRR under particular hypothesis on the coupling between the waveguide and the ring realizes an APF that induces a 360° steep phase shift across its resonance wavelength. If the optical carrier is placed close to the MRR resonance, slightly changing the reciprocal position of the carrier and the microring will change the phase of the carrier without affecting the sideband (Figure 3.16(b)) [73,74]. These approaches can take advantage of the fast phase control ensured by the photonic techniques, so that phase tuning time faster than 1 ns can be achieved, regardless of the RF carrier frequency that can easily be as high as several tens of gigahertz.

On the other hand, effective TTD based on photonics has been demonstrated exploiting several different approaches. The most straightforward one takes advantage of the low loss of optical waveguides to implement variable delays through optical path switching (Figure 3.17(a)): the signal is switched among few optical paths of different length, therefore with different propagation times [75].

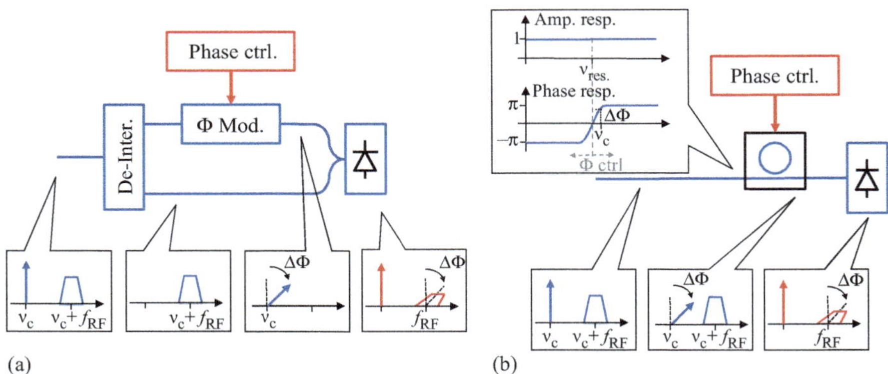

Figure 3.16 (a) Phase shifting through de-interleaving and phase control; (b) phase shifting through a wavelength-specific phase shifter, as a microring resonator in all-pass configuration

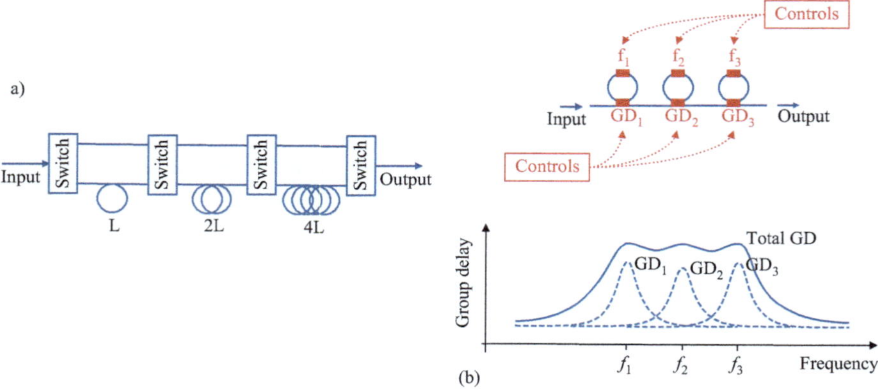

Figure 3.17 (a) Photonics-based TTD based on path switching; (b) principle of wideband TTD based on group delay (GD) synthesis by MRRs

A second type of TTD exploits the wavelength dependence of the laser propagation time through a medium due to the chromatic dispersion [75–78]: considering a modulated laser transmitted through, for example, an optical fiber, if the laser wavelength is changed, the propagation time in the dispersive element changes as well. Therefore, it is possible to control the delay of an optical signal by controlling its wavelength. If the RF signal to each antenna element is loaded on an independent laser, by tuning the lasers it is possible to control the time of emission of the RF signal of every element in the array. For example, the dispersion-compensating fiber has a dispersion of about –100 ps/nm·km; therefore, if a spool of 1 km is available, tuning the carrier wavelength of 1 nm will produce a variation in the time of arrival equal to 100 ps.

Another relevant method for realizing a TTD using photonics exploits the so-called slow light effect, that is, emulating the TTD by controlling the group delay of the optical sideband only [79]. This approach is based on the concept that the delay is the wavelength derivative of the phase; therefore, synthesizing phase variations with tunable steepness in the spectrum, it is possible to control the delay of the optical signal. An elegant implementation of this method has exploited the group delay of few cascaded optical micro-rings [80], controlling their resonance frequency through a phase control in the ring, and their group delay through the coupling between ring and waveguide (Figure 3.17(b)). This last example is particularly significant since it has demonstrated a photonic-integrated beamforming network based on TTD for broadband signals at high frequency.

The optical TTD is therefore an extremely interesting solution for high-performing beamforming. Nevertheless, its implementation is practically complex, in particular if tunable TTD is realized in integrated photonics, as in [80], where several parameters must be controlled simultaneously at each antenna element. Therefore, if the target application does not make use of ultra-wideband signals, an approach based on optical phase shifting can be more convenient, trading off a limited amount of squinting with a significant simplification in the operational complexity [72].

3.8 On-chip implementation: state of the art, future trends, and perspective

In order to reduce the SWaP of photonics-based subsystems, as well as to increase their reliability and the compliance to harsh environments, implementations on chip based on integrated photonics are needed. Silicon, III–V compounds, and glasses technologies are the most mature platforms to date for fabricating photonic integrated circuits (PICs). Each of the photonic subsystems discussed here can be optimized using the most suitable technology, but unfortunately, any of the available technological platforms is suitable for all of them. This suggests the hybrid integration approach, connecting together PICs of different platforms, as the most viable solution for the fabrication of entire photonics-based microwave systems.

Nevertheless, few components are still very critical, in particular considering the demanding radar or EW applications. Tunable optical filters with very high-quality factors for mode selection and sideband/carrier filtering, highly linear phase- and amplitude-modulators for electrical-to-optical conversion, low-noise and high-responsivity photodetectors for optical-to-electrical conversion are examples of fundamental devices the performance of which strongly affects the behavior of the entire microwave subsystems, and must therefore be carefully realized. Flat-top, abrupt-slope, ultra-narrow optical filters based on silicon technology have been recently reported using ring resonators with bandwidth in the order of few gigahertz or below, and out-of-band rejection exceeding 40 dB [81–83]. Efficient modulators can be realized by electro-absorption in the indium phosphide (InP) platform, showing significantly low driving voltages (1 V peak-to-peak [84]), an important factor for containing the power consumption. At the same time, InP-based MZMs have been improved considerably in recent years [85–87] and have become commercially available with qualified high performance. An example of photodetector for high-power applications is the uni-traveling-carrier (UTC)-type PD [88], whereas recently a new UTC structure with 1.5 μm thick p-InGaAs absorption layer has been proposed for high responsivity [89].

3.9 Summary

In this chapter, we have briefly analyzed the main concepts and functionalities of microwave photonics: the generation, detection, and distribution of microwave signals, the filtering of RF signals, and the control of their phase or time delay for the beamforming function in PAA.

From the presented schemes and results, it should be clear that photonics does have great potentials for transforming the microwave field. In particular, the frequency flexibility and the precision brought about by photonics can be exploited in microwave systems to reach new performance and new functionalities.

These aspects will be the main topic of Chapters 4–7.

References

[1] T. Debatty, "Software defined radar: a state of the art", 2nd International Workshop on Cognitive Information Processing, Elba, Italy, 2010.

[2] M.L. Skolnik, "Introduction to radar systems", 2nd Ed., McGraw-Hill, New York, 1980.

[3] M. Richards, J.A. Scheer, and W. A. Holm, "Principle of Modern Radar: basic principle", Raleigh, NC, USA: SciTech Publishing, 2010.

[4] J.B. Tsui, "Digital techniques for wideband receivers", 2nd Ed, Raleigh, NC, USA: SciTech Publishing, 2004.

[5] R. Walden, "Analog-to-digital conversion in the early twenty-first century", Wiley Encyclopedia of Computer Science and Engineering, 2008.

[6] A.J. Seeds, C.H. Lee, and M. Naganuma, "Guest editorial: microwave photonics". IEEE Journal of Lightwave Technology, vol. 21, no 12, pp. 2959–2960, 2003.

[7] C.H. Cox, and E.I. Ackerman, "Microwave Photonics: past, present and future", International Topical meeting on Microwave Photonics, pp. 9–11, 2008.

[8] J. Capmany, and D. Novak, "Microwave Photonics combines two worlds". Nature Photonics, vol. 1, no. 6, pp. 319–330. 2007.

[9] J. Yao, "Microwave photonics". IEEE Journal of Lightwave Technology, vol. 27, no. 22, pp. 314–335, 2009.

[10] J. Yu, Z. Jia, T. Wang, and G.K. Chang, "Centralized lightwave radio-over-fiber system with photonic frequency quadrupling for high-frequency millimeter-wave generation", IEEE Photonics Technology Letters, vol. 19, no. 19, pp. 1499–1501, 2007.

[11] T. Healy, F.C. Garcia Gunning, A.D. Ellis, and J.D. Bull, "Multi-wavelength source using low drive-voltage amplitude modulators for optical communications". Optics Express, vol. 15, no. 6, pp. 2981–2986, 2007.

[12] A.J. Metcalf, V. Torres-Company, and D.A.M. Weiner, "High-power broadly tunable electrooptic frequency comb generator". IEEE Journal of Selected Topics in Quantum Electronics vol. 19, no. 6, 2013.

[13] V. Torres-Company, and A.M. Weiner, "Optical frequency comb technology for ultra-broadband radio-frequency photonics". Laser and Photonics Reviews, 8, 3, 2013.

[14] R. Wu, V.R. Supradeepa, C.M. Long, D.E. Leaird, and A.M. Weiner, "Generation of very flat optical frequency combs from continuous-wave lasers using cascaded intensity and phase modulators driven by tailored radio frequency waveforms". Optics Letters, vol. 35, no. 19, pp. 3234–3236, 2010.

[15] D. Onori, F. Scotti, F. Laghezza, *et al.*, "A photonically-enabled compact 0.5 28.5 GHz RF scanning receiver". Journal of Lightwave Technology., 10.1109/JLT.
2018.2792304

[16] C.J. Buczek, R.J. Freiberg, and M.L. Skolnick, "Laser injection locking". Proceedings of the IEEE, vol. 61, no. 10, 1973.

[17] S. Pan, and J. Yao, "Wideband and frequency-tunable microwave generation using an optoelectronic oscillator incorporating a Fabry–Perot laser diode with external optical injection". Optics Letters, vol. 35, no. 11, pp. 1911–1913, 2010.

[18] L. Maleki, "The optoelectronic oscillator". Nature Photonics, vol. 5, no. 12, pp. 728–730, 2011.

[19] E. Rubiola, "The Leeson effect: PM and AM noise and frequency stability in oscillators, including OEOs and lasers". European Frequency and Time Forum (EFTF), 2014. IEEE, 2014.

[20] X.S. Yao, and L. Maleki, "High frequency optical subcarrier generator". Electronic Letters, vol. 30, no. 18, pp. 1525–1526, 1994.

[21] X. S. Yao, and L. Maleki, "Optoelectronic oscillator for photonic systems". IEEE Journal of Quantum Electronics, vol. 32, no. 7, pp.1141–1149, 1996.

[22] X. S. Yao, and L. Maleki, "Converting light into spectrally pure microwave oscillation". Optics Letters, vol. 21, no. 7, pp. 483–485, 1996.

[23] X. S. Yao, and L. Maleki, "Optoelectronic microwave oscillator". Journal of the Optical Society of America B, vol. 13, no. 8, pp. 1725–1735, 1996.

[24] V.S. Ilchenko, X.S. Yao, and L. Maleki, "High-Q microsphere cavity for laser stabilization and optoelectronic microwave oscillator". In Laser Resonators II, Proceedings of SPIE, San Jose, CA, USA, pp. 190–198, January 1999.

[25] L. Maleki, V. Iltchenko, S. Huang, and A. Savchenkov, "Micro optical resonators and applications in optoelectronic oscillators". In Proceedings of the IEEE International Topical Meeting on Microwave Photonics (MWP '02), Long Beach, CA, USA, January 2002.

[26] A.B. Matsko, L. Maleki, A.A. Savchenkov, and V.S. Ilchenko, "Whispering gallery mode based optoelectronic microwave oscillator". Journal of Modern Optics, vol. 50, no. 15–17, pp. 2523– 2542, 2003.

[27] A.A. Savchenkov, V.S. Ilchenko, J. Byrd, *et al.*, "Whispering gallery mode based opto-electronic oscillators". In Proceedings of the IEEE International Frequency Control Symposium (FCS '10), Newport Beach, CA, USA, pp. 554–557, June 2010.

[28] W. Loh, S. Yegnanarayanan, J. Klamkin, *et al*, "Amplifier-free slab-coupled optical waveguide optoelectronic oscillator systems". Optic Express, vol. 20, no. 17, pp. 19589–19598, 2012.

[29] X.S. Yao, L. Maleki, Y. Ji, G. Lutes, and M. Tu, "Dual-loop optoelectronic oscillator". In Proceedings of the IEEE International Frequency Control Symposium (FCS '98), Pasadena, CA, USA, pp. 545–549, May 1998.

[30] X.S. Yao, and L. Maleki, "Multiloop optoelectronic oscillator". IEEE Journal of Quantum Electronics, vol. 36, no. 1, pp. 79–84, 2000.

[31] E. Shumakher, and G. Eisenstein, "A novel multiloop optoelectronic oscillator". IEEE Photonics Technology Letters, vol. 20, no. 22, pp. 1881–1883, 2008.

[32] T. Bánky, B. Horváth, and T. Berceli, "Optimum configuration of multiloop optoelectronic oscillators". Journal of the Optical Society of America B, vol. 23, no. 7, pp. 1371–1380, 2006.

[33] W. Zhou and G. Blasche, "Injection-locked dual opto-electronic oscillator with ultra-low phase noise and ultra-low spurious level" IEEE Transactions on Microwave Theory and Techniques, vol. 53, no. 3, 2005.

[34] E. Levy, M. Horowitz, O. Okusaga, C. Menyuk, G. Carter, and W. Zhou, "Study of dual-loop optoelectronic oscillators". IEEE International Frequency Control Symposium, 2009 Joint with the 22nd European Frequency and Time forum.

[35] X.S. Yao, L. Davis, and L. Maleki, "Coupled optoelectronic oscillators for generating both RF signal and optical pulses". Journal of Lightwave Technology, vol. 18, no. I, 2000.

[36] X.S. Yao, L. Maleki, and L. Davis, "Coupled opto-electronic oscillators". 1998 IEEE International Frequency Control Symposium.

[37] J. Lasri, A. Bilenca, D. Dahan, *et al.*, "A self-starting hybrid optoelectronic oscillator generating ultra low jitter 10-GHz optical pulses and low phase noise electrical signals". IEEE Photonics Technology Letters, vol. 14, no. 7, 2002.

[38] J. Lasri, P. Devgan, R. Tang, and P. Kumar. "Self-starting optoelectronic oscillator for generating ultra-low-jitter high-rate (10 GHz or higher) optical pulses". Optic Express, vol. 11, no. 12, p. 1430, 2003.

[39] U. Kellers, "Recent developments in compact ultrafast lasers". Nature, vol. 424, pp. 831–838, 2003.

[40] T. Udem, R. Holzwarth, and T.W. Hänsch. "Optical frequency metrology". Nature, vol. 416, pp. 233–237, 2002.

[41] H.A. Haus, "Mode-locking of lasers". IEEE Journal of Selected Topics in Quantum Electronics, vol. 6, no. 6, 2000.

[42] H. Chi, and J.P. Yao, "An approach to photonic generation of high frequency phase-coded RF pulses". IEEE Photonics Technology Letters, vol. 19, no. 10, pp. 768–770, 2007.

[43] I.S. Lin, J.D. McKinney, and A.M. Weiner, "Photonic synthesis of broadband microwave arbitrary waveform applicable to ultrawideband communication". IEEE Microwave and Wireless Components Letters, vol. 15, no. 4, pp. 226–228, 2005.

[44] J. A. Nanzer, P.T. Callahan, M.L. Dennis, and T.R. Clark Jr. "Photonic signal generation for millimeter-wave communications", Johns Hopkins APL Technical Digest, vol. 30, no. 4, pp. 299–308, 2012.

[45] T. Yilmaz, C.M. DePriest, T. Turpin, J.H. Abeles, and P.J. Delfyett, "Toward a photonic arbitrary waveform generator using a modelocked

external cavity semiconductor laser". IEEE Photonics Technology Letters, vol. 14, no. 11, pp. 1608–1610, 2002.

[46] J. Chou, Y. Han, and B. Jalali, "Adaptive RF-photonic arbitrary waveform generator". International Topical Meeting on Microwave Photonics, pp. 1226–1229, 2002.

[47] G. Serafino, P. Ghelfi, P. Perez-Millan, *et al.*, "Phase and amplitude stability of EHF-band radar carriers generated from an active mode-locked laser". IEEE Journal of Lightwave Technology, v. 29, no. 23, pp. 3551–3559, 2011.

[48] P. Ghelfi, F. Scotti, F. Laghezza, and A. Bogoni, "Phase coding of RF pulses in photonics-aided frequency-agile coherent radar systems". IEEE Journal of Quantum Technology, vol. 48, no. 9, pp. 1151–1157, 2012.

[49] P. Ghelfi, F. Scotti, F. Laghezza, and A. Bogoni, "Photonic generation of phase-modulated RF signals for pulse compression techniques in coherent radars". IEEE Journal of Lightwave Technology, vol. 30, no. 11, pp. 1638–1644, 2012.

[50] G.C. Valley, "Photonic analog-to-digital converters". Optics Express, vol. 15, no 5, pp. 1955–1982, 2007.

[51] P.W. Juodawlkis, J.C. Twichell, G.E. Betts, *et al.*, "Optically sampled analog-to-digital converters". IEEE Transactions on Microwave Theory and Techniques, vol. 49, no. 2, pp. 1840–1853, 2001.

[52] P.W. Juodawlkis, J.C. Twichell, G.E. Betts, *et al.*, "Optically sampled analog-to-digital converters". IEEE Transaction on Microwave Theory and Techniques, vol. 49 no. 10, pp. 1840–1853, 2001.

[53] C. Boheémond, T. Rampone, and A. Sharaiha, "Performances of a photonic microwave mixer based on cross-gain modulation in a semiconductor optical amplifier". IEEE Journal of Lightwave Technology, vol. 29, no. 16, pp. 2402–2409, 2011.

[54] F. Laghezza, F. Scotti, P. Ghelfi, and A. Bogoni, "Photonics-assisted multiband RF transceiver for wireless communications". IEEE Journal of Lightwave Technology, vol. 32, no. 16, pp. 2896–2904, 2014.

[55] P. Ghelfi, F. Laghezza, F. Scotti, D. Onori, and A. Bogoni, "Photonics for radars operating on multiple coherent bands". Journal of Lightwave Technology, vol. 33, no. 2, pp. 500–507, 2016.

[56] F. Laghezza, F. Scotti, P. Ghelfi, S. Pinna, G. Serafino, and A. Bogoni, "Jitter-limited photonic analog-to-digital converter with 7 effective bits for wideband radar applications". IEEE Radar Conference, Ottawa, ON: Canada, pp. 906–920, 2013.

[57] J. Chou, J. Conway, G. Sefler, G. Valley, and B. Jalali, "150 GS/s real-time oscilloscope using a photonic front end". International Topical Meeting on Microwave Photonics, pp. 35–38, 2008.

[58] C.H. Cox, E.I. Ackerman, and J.L. Prince, "Limits on the performance of RF-over-fiber links and their impact on device design". IEEE Transactions on Microwave Theory and Techniques, vol. 54, no. 2, 2006.

[59] J. Beas, G. Castanon, I. Aldaya, A. Aragon-Zavala, and G. Campuzano, "Millimeter-wave frequency radio over fiber systems: a survey". IEEE Communications Surveys & Tutorials, vol. 15, no. 4, 2013.
[60] S. Futatsumori, K. Morioka, A. Kohmura, K. Okada, and N. Yonemoto, "Design and field feasibility evaluation of distributed-type 96 GHz FMCW millimeter-wave radar based on radio-over-fiber and optical frequency multiplier". Journal of Lightwave Technology, vol. 34, no. 20, 2016.
[61] J. Tao, C. Hong, Y. Gu, and A. Liu, "Demonstration of a compact wavelength tracker using a tunable silicon resonator". Optics Express, vol. 22, no. 20, pp. 24104–24110, 2014.
[62] J. Capmany, B. Ortega, and D. Pastor, "A tutorial on microwave photonic filters". IEEE Journal of Lightwave Technology, vol. 24, no. 1, pp. 201–229, 2006.
[63] B. Min, E. Ostby, V. Sorger, *et al.*, "High-Q surface-plasmon-polariton whispering-gallery microcavity". Nature, vol. 457, no. 7228, pp. 455–458, 2009.
[64] S.V. Ilchenko, M.L. Gorodetsky, X.S. Yao, and L. Maleki "Microtorus: a high-finesse microcavity with whispering-gallery modes". Optics Letters, vol. 26, no. 5, pp. 256–258, 2001.
[65] M.S. Rasras, D.M. Gill, S.S. Patel, *et al.*, "Demonstration of a fourth-order pole-zero optical filter integrated using CMOS processes". Journal of Lightwave Technology, vol. 25, no. 1, pp. 87–92, 2007.
[66] G. Serafino, C. Porzi, V. Sorianello, *et al.*, "Design and characterization of a photonic integrated circuit for beam forming in 5G wireless networks". Int. Top. Meet. Microwave Photonics (MWP) 2017, Mo1.4, Beijing, 2017.
[67] H.A. Macleod, "Thin-film optical filters," Boca Raton, FL, USA: CRC Press, 2001.
[68] X. Zou, M. Li, W. Pan, L. Yan, J. Azana, and J. Yao, "All-fiber optical filter with an ultra-narrow and rectangular spectral response". Optics Letters, vol. 38, no. 16, pp. 3096–3098, 2013.
[69] C. Porzi, G. Serafino, P. Velha, P. Ghelfi, and A. Bogoni, "Integrated SOI high-order phase shifted Bragg grating for microwave photonics signal processing". Journal of Lightwave Technology, vol. 35, no. 20, pp 4479–4487, 2017.
[70] H. Shen, M.H. Khan, L. Fan, *et al.*, "Eight-channel reconfigurable microring filters with tunable frequency, extinction ratio and bandwidth". Optics Express, vol. 18, no. 17, pp. 18067–18076, 2010.
[71] G. Serafino, C. Porzi, V. Sorianello, *et al.*, "Design and characterization of a photonic integrated circuit for beam forming in 5G wireless networks". MWP 2017.
[72] F. Falconi, C. Porzi, S. Pinna, *et al.*, "Fast and linear photonic integrated microwave phase-shifter for 5G beam-steering applications". OFC 2018.
[73] M. Pu, L. Liu, W. Xue, *et al.*, "Widely tunable microwave phase shifter based on silicon-on-insulator dual-microring resonator". Optics Express, vol. 18, no. 6, pp. 6172–6182, 2010.

[74] D.B. Adams, and C.K. Madsen, "A novel broadband photonic RF phase shifter". Journal of Lightwave Technology, vol. 26, no. 15, pp. 2712–2717, 2008.

[75] R. Soref, "Optical dispersion technique for time-delay beam steering". Applied Optics, vol. 31, no. 35, pp. 7395–7397, 1992.

[76] L. Yaron, R. Rotman, S. Zach, and M. Tur, "Photonic beamformer receiver with multiple beam capabilities". IEEE Photonics Technology Letters, vol. 22, no. 23, pp. 1723–1725, 2010.

[77] K. Prince, M. Presi, A. Chiuchiarelli, *et al.*, "Variable delay with directly-modulated R-SOA and optical filters for adaptive antenna radio-fiber access". IEEE Journal of Lightwave Technology, vol. 27, no. 22, pp. 5056–5064, 2009.

[78] F. Scotti, P. Ghelfi, F. Laghezza, G. Serafino, S. Pinna, and A. Bogoni, "Flexible true-time-delay beamforming in a photonics-based RF broadband signals generator". IET Conference Proceedings, London, UK, pp. 789–791, 2013.

[79] A. Zadok, O. Raz, A. Eyal, and M. Tur, "Optically controlled low-distortion delay of GHz-wide radio-frequency signals using slow light in fibers". IEEE Photonics Technology Letters, vol. 19, no. 7, pp. 462–464, 2007.

[80] A. Meijerink, C.G. Roeloffzen, R. Meijerink, *et al.*, "Novel ring resonator-based integrated photonic beamformer for broadband phased array receive antennas—part i: design and performance analysis". IEEE Journal of Lightwave Technology, vol. 28, no. 1, pp. 3–18, 2010.

[81] P. Alipour, A.A. Eftekhar, A.H. Atabaki, *et al.*, "Fully reconfigurable compact RF photonic filters using high-Q silicon microdisk resonators". Optics Express, vol. 19, no. 17, pp. 15899–15907, 2011.

[82] P. Dong, N.N. Feng, D. Feng, *et al.*, "GHz-bandwidth optical filters based on high-order silicon ring resonators". Optics Express, vol. 18, no. 23, pp. 23784–23789, 2010.

[83] M.S. Rasras, K.Y. Tu, D.M. Gill, *et al.*, "Demonstration of a tunable microwave-photonic notch filter using low-loss silicon ring resonators". IEEE Journal of Lightwave Technology, vol. 27, no. 12, pp. 2105–2110, 2009.

[84] J.W. Raring, L.A. Johansson, E.J. Skogen, *et al.*, "40-Gb/s widely tunable low-drive-voltage electroabsorption-modulated transmitters". IEEE Journal of Lightwave Technology, vol. 25, no. 1, pp 239–248, 2007.

[85] K. Prosyk, A. Ait-Ouali, C. Bornholdt, *et al.*, "High performance 40 GHz InP Mach-Zehnder modulator". Optical Fiber Communication Conference (OFC), Los Angeles, USA, March 4–8, 2012.

[86] K.-O. Velthaus, M. Hamacher, M. Gruner, *et al.*, "High performance InP-based Mach-Zehnder modulators for 10 to 100 Gb/s optical fiber transmission systems". 23rd International Conference on Indium Phosphide and Related Materials (IPRM), Berlin, Germany, May 22–26, 2011.

[87] K. Prosyk, A. Ait-Ouali, J. Chen, *et al.*, "Travelling wave Mach-Zehnder modulators". 25rd International Conference on Indium Phosphide and Related Materials (IPRM), Kobe, Japan, May 19–23, 2013.

[88] H. Ito, S. Kodama, Y. Muramoto, T. Furuta, T. Nagatsuma, and T. Ishibashi, "High-speed and high-output InP-InGaAs unitraveling-carrier photodiodes," IEEE Journal Selected Topics in Quantum Electronics, vol. 10, no. 4, pp. 709–727, 2004.
[89] M. Chtioui, F. Lelarge, A. Enard, *et al.*, "High responsivity and high power UTC and MUTC GaInAs-InP photodiodes". IEEE Photonics Technology Letters, vol. 24, no. 4, pp. 318–320, 2012.

Chapter 4
Photonics-based radars

Filippo Scotti[1], *Paolo Ghelfi*[1], *and Antonella Bogoni*[1,2]

4.1 Chapter organization and key points

This chapter describes how the basic concepts of microwave photonics, as introduced in Chapter 1, can be combined to implement complete photonics-based radar systems.

The chapter first describes the possible architectures of a photonics-based radar transceiver and reports the most significant examples of implementations of complete radar systems. Then, the even more peculiar case of a multiband radar transceiver enabled by the photonic approach is described, together with an analysis of the specific processing required by such a system, and with a set of relevant case studies and implementations of dual-band photonics-based radars. As a confirmation of the flexibility of the multiband radar system enabled by the use of photonics, the reported examples of dual-band radar deal with the maritime and the aerial scenarios, and the remote sub-millimeter displacement measurement. In these analyses, comparisons with conventional systems are also provided, in particular in the case of the single-band radar for maritime application, where a significant direct comparison is reported.

4.2 Photonics-based transceivers

The microwave photonic functionalities described in Chapter 3 can be combined to realize a complete photonics-based radio frequency (RF) transceiver that can be used in applications such as radar or communication systems, taking advantages of the peculiar features brought about by photonics.

Rather than independently exploiting a transmitter and a receiver both implemented by photonic techniques, a photonics-based transceiver can be implemented that share the laser source between each subsystem, conveniently reducing the size of the transceiver and the cost of the photonics-based approach, as sketched in Figure 4.1.

[1]National Laboratory of Photonic Networks and Technology (PNTLab), Consorzio Nazionale Interuniversitario per le Telecomunicazioni (CNIT), Italy
[2]Institute of Technologies for Communication, Information and Perception (TeCIP), Sant'Anna School of Advanced Studies, Italy

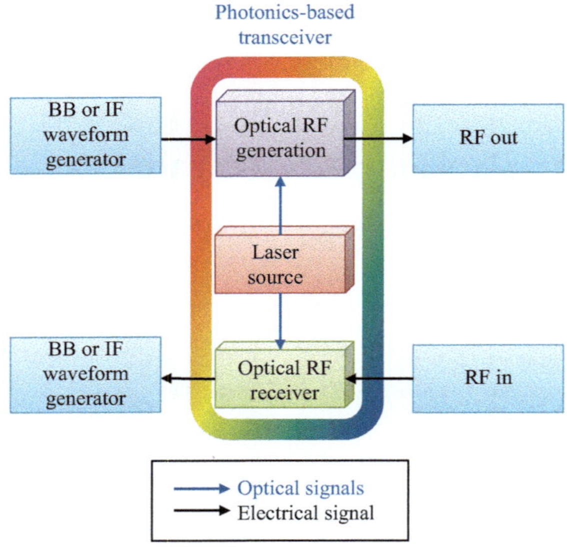

Figure 4.1 Scheme of the photonics-based transceiver

Moreover, as highlighted in Chapter 3, several additional functionalities can be implemented directly in the photonic domain. While a lot of work has been done for developing each single photonics-based functionalities, only a few examples of complete photonics-based transceivers have been presented so far, in particular related to radar applications [1,2] and to wireless communication applications [3,4].

Here, we focus on the exploitation of the concept of the photonics-based transceiver to radar systems, reporting two recent examples of photonics-based transceivers for radar applications.

4.3 A general photonics-based transceiver for software-defined radars

In the seminal work reported in [1], a mode locked laser (MLL) with a repetition rate of 400 MHz has been used as the master clock of the photonic transceiver governing the radar system.

Concerning the generation of the RF signal, the solution adopted in [1] is slightly different from the scheme reported in Figure 3.2(b) in Chapter 3. In particular, the function of the optical filters (not available with the selectivity required by the repetition rate of the exploited MLL) has been carried out by RF filters after the photodiode (PD), with the additional advantage of avoiding the phase instabilities brought about by an interferometric structure realized with discrete components (see the scheme of Figure 3.2(b)).

The radar waveform has been digitally generated at a low intermediate frequency (IF) and fed into the RF port of an electro-optical modulator to modulate

Table 4.1 Performance characterization of the photonics-based transceiver

Parameter	Photonics-based transceiver	State of the art electronics transceiver
	Transmitter	
Carrier frequency	Flexible direct generation up to 40 GHz	Direct generation below 2 GHz up-conversions above 2 GHz
Signal jitter	<15 fs integrated in (10 kHz to 10 MHz)	Typical >20 fs integrated in (10 kHz to 10 MHz]
Signal-to-noise ratio (SNR)	>73 dB/MHz	>80 dB/MHz
Spurious-free dynamic range (SFDR)	>70 dBc	>70 dBc
Instantaneous bandwidth	200 MHz, easily extendable with MLL at higher repetition rate	<2 GHz
	Receiver	
Input carrier frequency	Up to 40 GHz with direct RF undersampling	<2 GHz down-conversions at higher frequencies
Instantaneous bandwidth	200 MHz, easily extendable with MLL at higher repetition rate	<2 GHz
Sampling jitter	<10 fs integrated in (10 kHz to 10 MHz)	Typical >100 fs integrated in (10 kHz to 10 MHz)
Spurious-free dynamic range (SFDR)	50 dB	>70 dB
Effective number of bits (ENOB)	>7 for carrier frequency up to 40 GHz	<8 for carrier frequency <2 GHz

the MLL output. This optical signal is then detected by a PD. The beating products generated by the detection process in the PD generate a replica of the radar waveform at every frequencies equal to IF plus an integer multiple of the MLL repetition rate. Then, as said, an RF filter at the output of the PD selects the proper signal replica to be transmitted.

Concerning the receiver, the detected RF signal drives another electro-optical modulator, as reported in Figure 3.3(b) in Chapter 3. The optical signal from the MLL is amplitude-modulated by the received RF signal, and then photodetected by means of a low bandwidth PD, down-converting the detected RF signal (the radar echo) from the RF carrier back to the original IF.

The performance of the photonics-based transceiver presented in [1] is reported in Table 4.1 compared with the electronic transceivers at the state of the art. As can be seen from the table, the photonic approach shows advantages in the extreme frequency flexibility over tens of gigahertz, and in the precision of the digitization for any input frequency, enabling the software-defined radio paradigm.

4.4 A specific photonics-based transceiver for FMCW radar

The work reported in [2], instead, has been specifically implemented to realize a particular radar scheme known as linear frequency modulated continuous wave (FMCW). This scheme considers the emission of a continuous wave (CW) radar signal that is always on (no pulsed behavior), and linearly changes the frequency of the emitted CW with time. To realize this modulation scheme, the cited work generates two optical signals from a single laser by modulating it with an external RF signal changes its frequency linearly with time. The photonics-based scheme exploits a particular setting of the optical modulator that makes use of the modulator nonlinearity to realize a frequency multiplication by a factor 4, thus relaxing the requirements on the digital signal generation. This RF generation scheme has been described in Section 3.3.

The optical signal source described above is used also for receiving the radar echo, directly implementing the so-called de-chirping of the linear FMCW signal: in fact, the frequency difference between the transmitted radar signal and the received echo gives information about the distance of the target.

Although less flexible than the scheme presented in [1], this photonics-based implementation allows to obtain a broadband radar signal. In fact, the work in [2] has shown a linear FMCW across a bandwidth of 8 GHz. Moreover, as in the case above, the signal can be digitized by a simple analog-to-digital converter (ADC) at 100 MSample/s, thanks to the de-chirping mechanism.

4.5 Photonics-based radars and field trials

The photonics-based transceivers described above have been exploited for implementing coherent radar systems.

The architecture reported in [1] has been developed in a radar prototype named PHODIR (photonics-based fully digital radar), from the name of the original project that funded its development.

The block diagram of the PHODIR prototype is reported in Figure 4.2, showing a front-end in the X-band (namely, at 9.9 GHz) connected to the photonics-based transceiver. Figure 4.2 also reports a picture of the developed prototype.

The PHODIR demonstrator has been tested in several field trial experiments in different scenarios, evaluating the effectiveness of the photonic approach inside a real radar system [5,6].

Here, it is worth reporting on the field trial that has been run in collaboration with a company producing radar systems for maritime navigation since it has allowed a direct comparison with a state-of-the-art commercial system (named "SeaEagle"), working in the X-band as well, using the same kind of RF signals [6].

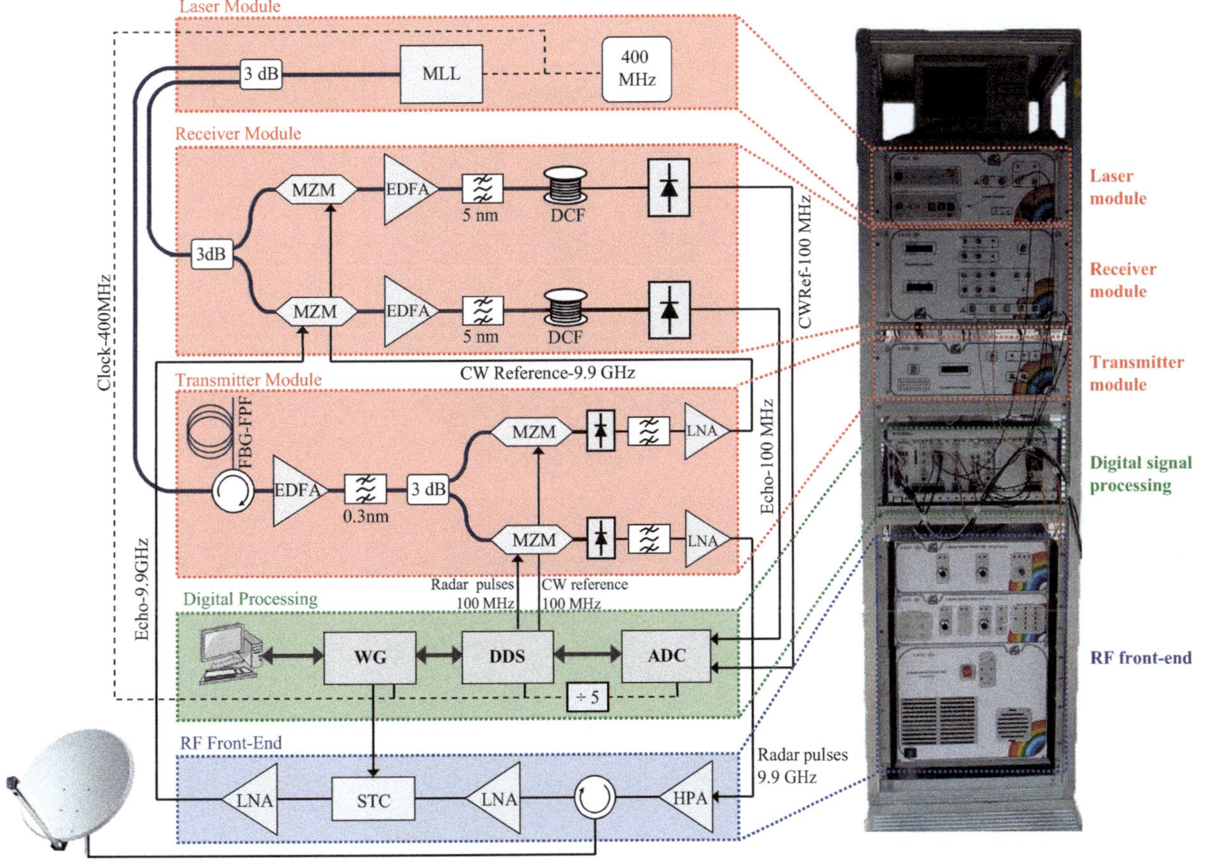

Figure 4.2 Block diagram of the PHODIR prototype and a picture of the demonstrator. Thick lines: optical path; thin lines: electrical path; dashed lines: clock connections

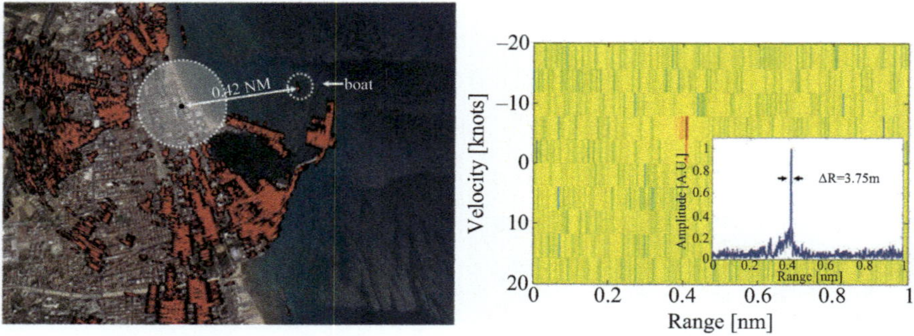

Figure 4.3 PHODIR radar detection of the San Benedetto del Tronto harbor area (left). Range-Doppler map of the detected boat (right). Inset: range profile of the detected target

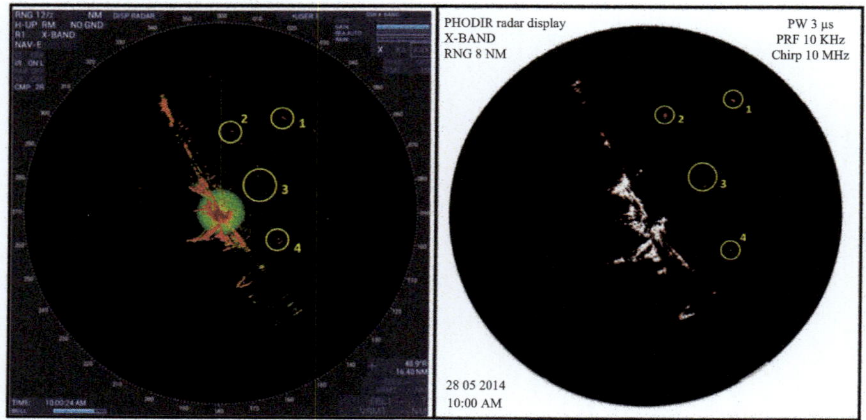

Figure 4.4 Comparison of a commercial radar (left) and of the photonic single-band radar (right)

Figure 4.3 (left) shows the detection trace taken by the photonics-based radar. The coastal area and harbor shape are well represented, and a small echo from a moving boat at a distance of about 0.42 nautical miles (NM) is also detected. The range/velocity map of this scene is reported in Figure 4.3(right), clearly identifying the small boat at 0.42 NM approaching with a speed of 5 knots (the negative sign indicates the direction for reducing distance). The inset in Figure 4.3(right) reports the range profile including the detection of the boat, highlighting a radar range resolution of 3.75 m, given by the employed chirped pulses with a bandwidth of 40 MHz.

Figure 4.4 depicts the detection comparison between the PHODIR demonstrator (right) and the commercial radar system. The graph shows several target detections

Table 4.2 Quantitative performance comparison between the photonics-based radar (PHODIR) and the commercial maritime radar (SeaEagle)

Parameters	**SeaEagle**	**PHODIR**
Peak power	100 W (@ WR90)	50 W (@ WR90)
RF (MHz)	9300 ÷ 9500 (step)	9880 ÷ 9920 (continuous)
Max bandwidth	40 MHz	40 MHz
Noise figure	5.5 dB	8 dB
MDS	–90 dBm @ 40 MHz BW	–87 dBm @ 40 MHz BW
Processing gain	Max 30 dB @ PW = 93 μs, B = 10 MHz (only compression gain)	31 dB @ T_D = 12.5 ms, PW = 5 μs, B = 2 MHz, PRF = 10 kHz (G_I = 21 dB + G_C = 10 dB)
Frequency accuracy	100 ppm (10 fs)	120 ppm (13 fs)
Max range	48 NM (cargo target)	18 NM (cargo target)
Pulsewidth	50 ns ÷ 93 μs	0.2 ÷ 10 μs
PRF	350 ÷ 2500 Hz	1 ÷ 12 kHz
Modulation format	Chirp	Any

within the maximum unambiguous range (8 NM), as well as the detection of the coastal area. It is important to underline that both the radars detected the all the targets, so they show similar performance even if the PHODIR radar is just a demonstrator, while the SeaEagle is a fully developed product. From this result, one can conclude that the use of photonics in a radar system is at least not detrimental.

Table 4.2 reports a more quantitative comparison between the two radars (the PHODIR prototype and the SeaEagle), reporting their main parameters.

The minimum detectable signal (MDS) of the photonics-assisted radar (–87 dBm) turns out to be comparable with the MDS of the commercial system (–90 dBm), even though the SeaEagle is capable of numerically suppressing the sea clutter thanks to advance processing capabilities (not reported in the table). This explains the large difference in the maximum range. On the other hand, the PHODIR transceiver is known to offer the peculiar potential features of bandwidth extension and frequency (in the table, limited by the RF filters selecting the beating at the PD), and waveform agility.

The photonics-based radar developed in [2] has been tested in a very interesting field trial as well.

In this case, the frequency quadruplicating scheme has allowed generating a linear FMCW signal in K-band with the considerable bandwidth of 8 GHz, from 18 GHz to 26 GHz, enabling a high-resolution radar detection. Moreover, the photonics-based de-chirping of the linear FMCW has permitted the use of a low-sampling rate ADC despite the huge signal bandwidth, allowing the implementation of the complex inverse synthetic aperture radar (ISAR) imaging technique.

The reported results show a two-dimensional imaging resolution of about 2 cm × 2 cm at 100 frames per second. This detection capability has been applied to the imaging of an electric fan while its blades were moving.

4.6 Multiband photonics-based transceiver and radar

As anticipated in Chapter 3, photonics allows generating and detecting multiple RF signals on different RF bands simultaneously. As discussed therein, while in conventional RF schemes this would require to duplicate all the transmitter and receiver circuits for each of the bands, in the photonics-based transmitters and receivers most of the optical devices in the schemes can be shared for managing the different bands.

Therefore, since multiband photonics-based transmitters and receivers are available, it is also possible to implement a multi-band RF transceiver. This is the approach followed in [7] (Figure 4.5) by the research group that developed the PHODIR prototype. In the cited work, the multiband RF generator and the multiband RF receiver are both fed by the same MLL. A digital waveform generator (direct digital synthesizer [DDS]) provides simultaneous multiple signals at different intermediate frequencies (IF_i), which are fed to the RF transmitter block. Here, the optical signal from the MLL is mixed with the electrical signals at IF, generating multiple RF signals at different frequency bands (as sketched in inset D of Figure 4.5). Then, multiple passband RF filters centered at different central frequencies (CF_i) select the desired RF signals.

Similarly, the RF receiver block combines the multiple received RF echoes with the optical signal from the MLL, obtaining the simultaneous down-conversion of each detected RF signal to its original IF (as sketched in inset F). Then, a single ADC digitizes all the detected signals simultaneously.

The scheme presented in [7] also includes a broadband front-end, constituted by a power amplifier at the transmitter side and a preamplifier at the receiver side. The broadband front-end guarantees the frequency flexibility of the multiband photonics-based transceiver. Nevertheless, an approach based on multiple frequency selective RF front-ends could optimize the system performance (e.g., in terms of noise figure, linearity, dynamic range, etc.) and could be therefore preferable for demanding applications.

It is also worth underlining that, in the scheme proposed in [7], since the RF signals are all generated and detected using the same MLL, they are all intrinsically phase locked to each other. Therefore, as will be detailed in the following, the scheme combines the capability for a multiband software-defined approach with the potential for a simple fusion of the data from multiple bands, and a consequent improvement of the system precision.

As reported in [7], the single-band system of the PHODIR prototype has been expanded with a second front-end in the S-band (namely, at 2.5 GHz) for dual-band functionality, thus exploiting the signal coherence provided by photonics. This new prototype has been named PANDORA (photonics-based dual-band radar).

The PANDORA transceiver has been characterized in a controlled environment [8] to check its performance. In this case, two separate front-ends have been used. The measurements have been carried out by connecting the RF outputs of the transceiver (at about –30 dBm peak power) to the antenna ports of the two RF

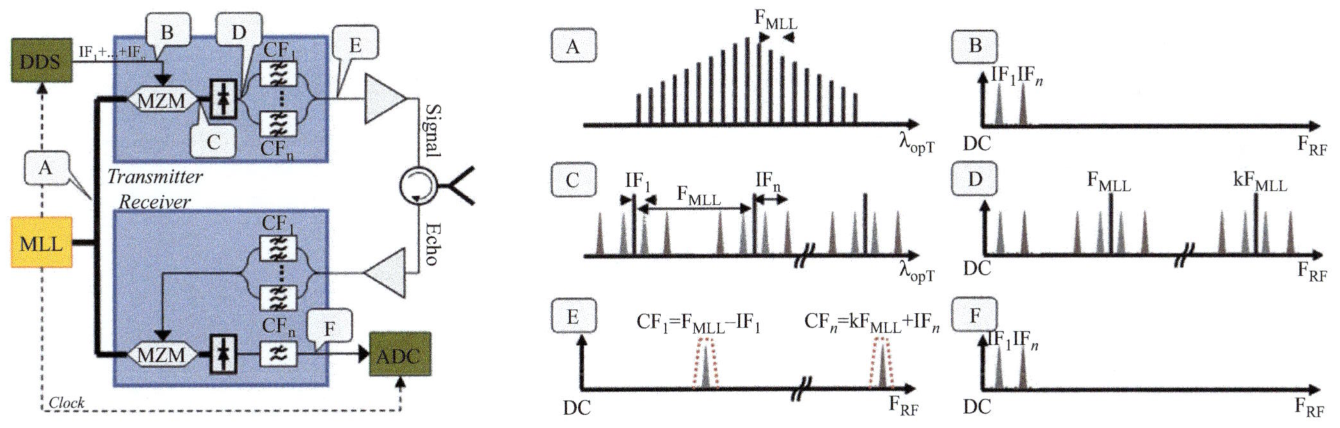

Figure 4.5 Architecture of a multiband photonics-based radar transceiver [7]

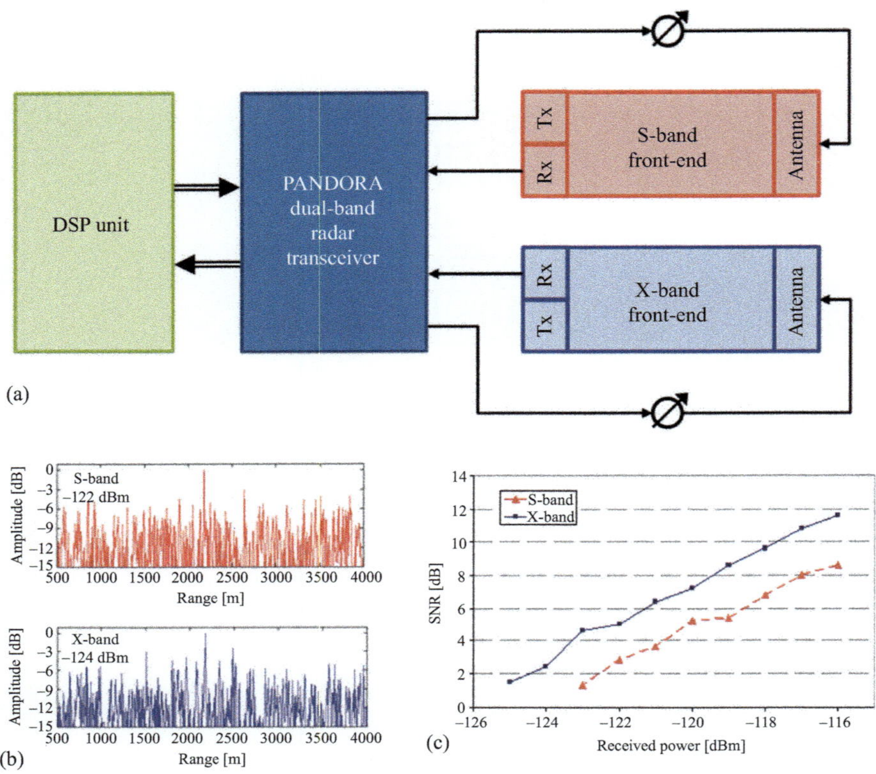

Figure 4.6 (a) Experimental setup for the system characterization; (b) range profiles of the target at the sensitivity levels; (c) measured SNR of the emulated target varying the RF power

front-ends via variable attenuators, as sketched in Figure 4.6(a), thus emulating the received echoes. The system has been set to generate 5 μs linearly chirped pulses with a bandwidth of 15 MHz and a pulse repetition frequency (PRF) of 10 kHz in both the S- and X-band. The RF powers have been reduced down to the sensitivity limit of the transceiver, with the target appearing just 3 dB above the noise floor. This situation is reported in Figure 4.6(b), which shows the range profiles corresponding to an echo power of –122 and –124 dBm in the S- and X-band, respectively. In both the frequency bands, the target appears at a distance of 2200 m due to an optical fiber spool delaying the signal inside the transceiver. Figure 4.6(c) reports the signal-to-noise ratio (SNR) of the peaks as detected by the system in the two frequency bands, varying the power at the front-ends inputs. The SNR here is considered as the ratio between the target peak and the strongest peak of the noise floor. The two curves linearly increase with the input power, and the SNR of the X-band target is about 2 dB higher than the one of the S-band, in accordance with the different performance of the two RF front-ends.

4.7 Dual-band signal processing

The intrinsic coherence of the PANDORA dual-band radar described above is a peculiar feature that implies significant advantages in managing the signals and their digital processing. In the following, these advantages are described and discussed.

The architecture of the dual-band signal processing developed in PANDORA is shown in Figure 4.7 [8,9]. As described in the previous paragraph, the architecture of the PANDORA dual-band radar allows detecting the two signals at different IF (coming from the S- and X- bands) with a single ADC. The ADC is

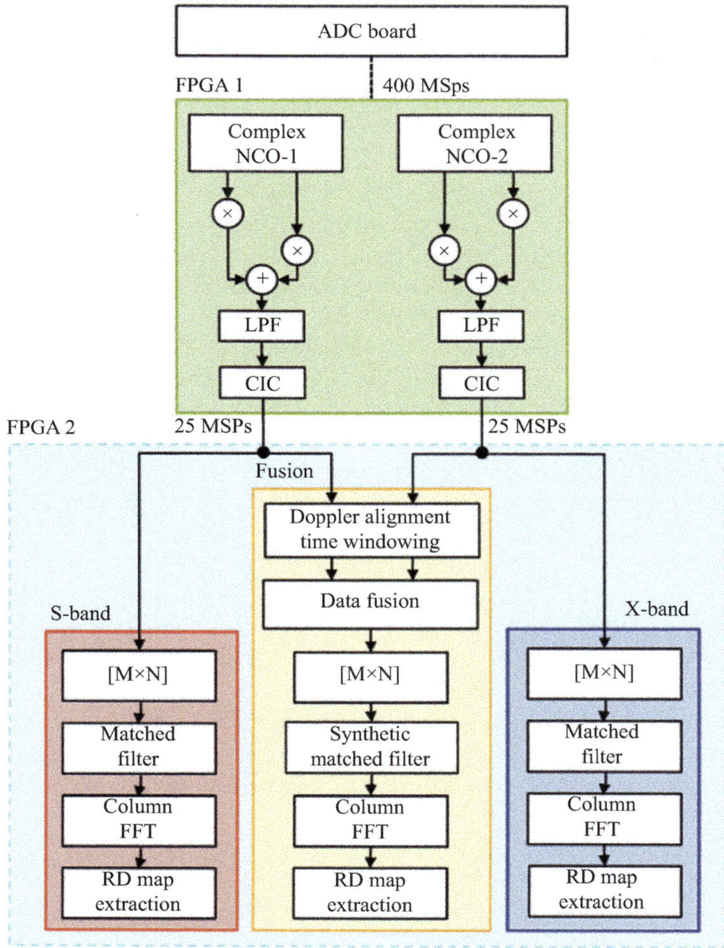

Figure 4.7 Block diagram of the dual-band processing through FPGAs

clocked at 400 MSps by the MLL. Since the MLL also clocks the DDS generating the radar pulses at IF, all the digital signals in the system are therefore perfectly synchronous, and the MLL works as master clock of the entire system.

The digital samples from the ADC are fed to a field programmable gate array (FPGA1), clocked to the MLL as well, that separates the two echoes and down-converts them to base-band. To this extent, the samples at the two IFs are multiplied by two complex numeric controlled oscillators (NCOs) at exactly the two IFs, so that each channel is moved to base-band as a complex signal. In this operation, the coherence of all the digital signals permits to construct the NCOs at the exact sampling rate of the detected echoes, so that the digital down-conversion is a mere multiplication. Then, each signal is precisely filtered exploiting an opportunely designed low pass filter (LPF). A typical digital LPF is a 61-taps finite impulse response (FIR) filter, with a cut-off frequency in accordance with the signal bandwidth. For example, if the signals at IF show a BW of 20 MHz, the appropriate filter should conveniently have a cut-off frequency of 10 MHz. Since the sampling rate of 400 MSps is no longer required at this stage, two cascaded integrated combs decimate the samples of the two channels. In the example case of baseband signals with 10 MHz BW, a decimation by a factor 16 can be used, reducing the sampling rate down to 25 MSps, thus strongly simplifying the processing and storage load. Before exiting the FPGA, the two data streams are temporary saved in two memory banks that act as buffers to prevent data losses, and then they can be moved in a hard disk to be processed offline, or sent to a second FPGA (FPGA2) to perform other calculations.

The most common calculation is the range/Doppler coherent processing (light blue block in Figure 4.7). In this case, the data are organized in $M \times N$ matrices, with M the number of radar acquisitions and N the samples per acquisition. The data thus represented are filtered with the matched filter required by the transmitted waveform. In this operation as well, the precise coherence of the digital signals makes the processing straightforward. Then a fast Fourier transform (FFT) along the matrix columns is calculated and coherently integrated across the M acquisitions. These operations allow the calculation of the range/velocity maps (RD: range/Doppler). Several other data processing are also possible, as thresholding, tracking, imaging, etc.

It is worth highlighting that the SDR approach enabled by the architecture of the PANDORA system allows changing the features of each signal (as pulse shape, duration, bandwidth, repetition rate, etc.) according to the observed scene. These changes require to accordingly modifying the matrices dimensions and the matched filters for each channel. This can be implemented by reconfiguring the processing parameters via software.

Besides the single-band processing, the coherent dual-band radar system also allows implementing a data fusion to get a better detection of the observed scene. To this extent, the intrinsic coherence of the two bands allows implementing a simplified fusion algorithm, based on the coarse approximation that two identical radar waveforms at different frequency bands are equally scattered by the observed targets, and are subject to equal losses during propagation. If this hypothesis can be accepted, the concatenation of two delayed chirped pulses with identical length and bandwidth produces a chirped pulse with double length and double bandwidth, thus doubling the range resolution as well. Therefore, only a Doppler shift correction and a time alignment are required (due to the different wavelength, see yellow block in Figure 4.7). Finally, the synthetic signal is calculated by simply summing up the two echoes. For comparison, the data fusion for noncoherent signals requires iterative algorithms for aligning the samples and permitting the calculation [10–12].

4.8 Case study: naval scenario field trial

To test the PANDORA system in a real environment, the group that implemented the dual-band radar organized a field trial campaign in collaboration with an Institute of the Italian Navy (the CSSN-ITE Institute in Livorno, Italy) [8], focusing on the maritime scenario in front of the coast of the city of Livorno. Figure 4.8 shows pictures of the PANDORA system installed at CSSN-ITE, highlighting the blocks constituting the dual-band radar, including the two front-ends in the X- and S-bands.

A first set of measures was conducted exploiting the naval traffic in front of the port of Livorno. The scenario is represented in Figure 4.9(a), reconstructed thanks to the automatic identification system (AIS) data received by the vessels present in the sea area. These vessels were a ferry ship approaching the port from southwest, and other ships anchored in the harbor within the radar line of sight. The PANDORA

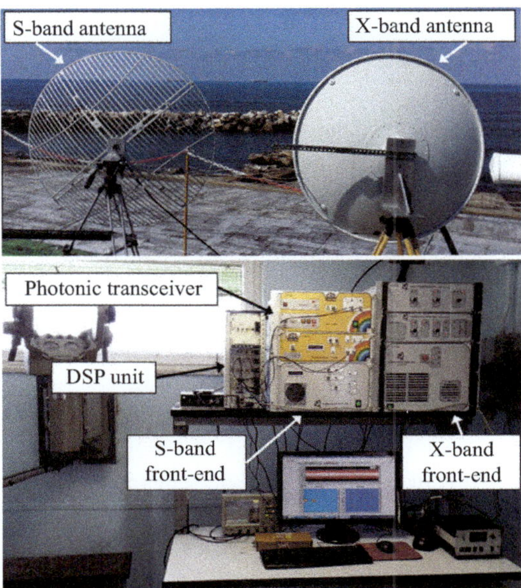

Figure 4.8 Picture of the system as installed at the CSSN-ITE institute

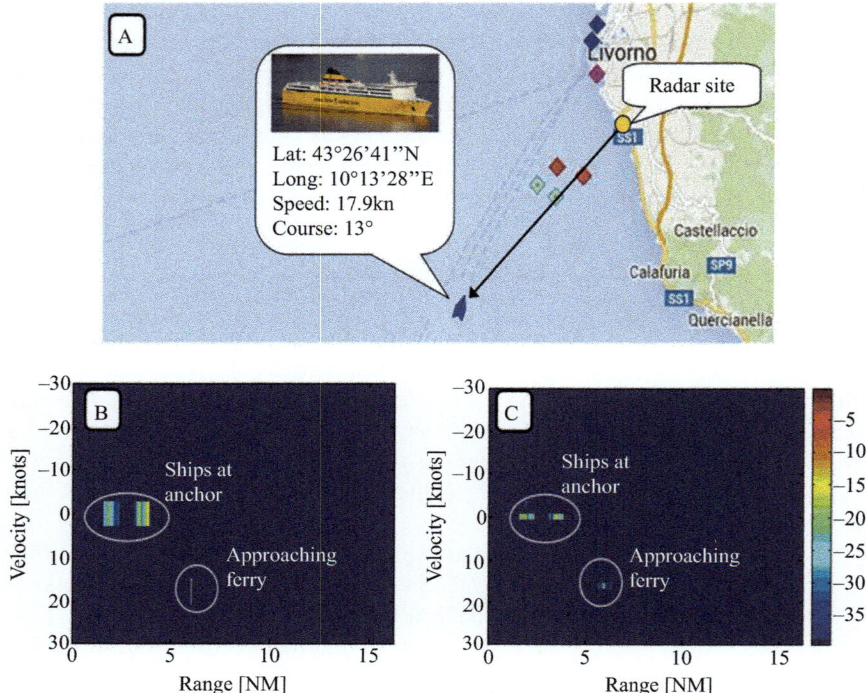

Figure 4.9 (a) Maritime scenario detected in front of the Livorno port, based on the AIS data; (b) range/Doppler map as detected in the S-band; (c) range/Doppler map as detected in the X-band

system was set with the same waveform for both the frequency bands: 5-μs-long radar pulses with a linear frequency chirp of 18 MHz, and with a PRF of 5 kHz, giving a range resolution of about 8 m with an unambiguous range of 16 NM. The range/velocity maps obtained in the S-band and in the X-band for the analyzed scenario are reported in Figure 4.9(b) and (c), respectively, in accordance with the AIS data.

It is interesting to note that the maps for the S- and X-bands differ in terms of the velocity resolution, which appears higher in the X-band. This is due to the fact that, for a given speed, the Doppler shift undergone by the radar signal is proportional to its carrier frequency. Since the minimum detectable frequency variation is about 50 Hz for a coherence integration time (CIT) of 20 ms, the velocity resolution turns out to be about 3 m/s in the S-band and 0.75 m/s in the X-band. Therefore, in the considered measurement set, the radar signal in the S-band estimated a cruise speed of about 18 kn, whereas the X-band signal provided a velocity evaluation of about 16 kn. Since the AIS data reported a cruise speed of 17.9 kn, when rescaling this value according to the radar viewing angle (about 25°), a radial velocity of 16.2 kn could be calculated.

As discussed above, the signal in the X-band is more precise in estimating the target velocity, as was actually expected. On the other hand, the signal in the S-band is known to have a longer covering range. A dual-band system can take advantage of both!

A second measurement campaign was run exploiting a 30-m-long cooperative boat provided by the CSSN-ITE, with the focus of verifying the accuracy of the photonic-based radar system and its performance in the two frequency bands.

The vessel was equipped with a GPS recorder, and it was made cruising from the coast to the open sea with constant trajectory and speed. The radar system was set as in the measurement reported above, with both the frequency bands generating a 5-μs-long linearly chirped pulse with a PRF of 5 kHz. Figure 4.10 shows the range/Doppler maps acquired in the X-band (a–c) and in the S-band (d–f), as well as the range profiles in linear scale (g–i), taken at different moments, that is,, when the cooperating vessel was at different distance from the port. The data in the first column were all taken at 14:10, and the target is clearly visible in both the maps at about 1.5 NM moving at about 7 kn. In the range profiles, a secondary peak can be appreciated in the X-band echo, given by a secondary scatterer of the boat (probably the aft cranes), which is not evident in the S-band. The data in the second column were taken at 14:35, when the target reached a distance of 4.3 NM from the coast. This is clear from the map related to the X-band, while it is less evident in the S-band due to the lower sensitivity of the measurement. This different sensitivity was ascribed to the different transmission powers and antenna gains of the front-ends, and to the frequency dependence of the radar cross section (RCS) of the target. Anyway, a proper decision algorithm can still easily identify the target from the S-band acquisition since its peak is 7 dB above the noise and clutter level. The maps at 14:35 also show a stationary target at 2.5 NM that corresponded to a cargo vessel that arrived in the meantime and was waiting to enter the port. Finally, the data reported in the third column were taken at 15:00: the cargo vessel was still there, the cooperative target is visible in the X-band even if it appeared only 3 dB above the noise/clutter floor, whereas it was no more detectable by the S-band.

Figure 4.11(a) reports the range of the target detected at different moments by the S- and X-band channels and compares it with the data recorded by the onboard GPS. The three traces are all in perfect agreement, confirming the proper operation of the photonics-based system. Similarly, Figure 4.11(b) reports the velocity of the target as detected by the two radar bands and by the GPS: again, the radar traces agree with the GPS data, taking into account the different resolution that depends on the duration of the measure (the CIT). Moreover, an accurate analysis of the RF power at the photonics-based radar receiver was realized, taking into account the frequency dependence of the target RCS [13] and the gain difference of the two front-ends. The result, reported in Figure 4.11(c), shows that the minimum RF power detected at the radar receiver (corresponding to the maximum detectable range) is equal for the X- and S-bands, confirming the frequency independence of the photonic transceiver.

Finally, the maritime field trial was also exploited to run an ISAR imaging experiment [14]. The PANDORA system was pointed at a container ship (Figure 4.12(a)) moving in front of the port of Livorno. The radar signals were both

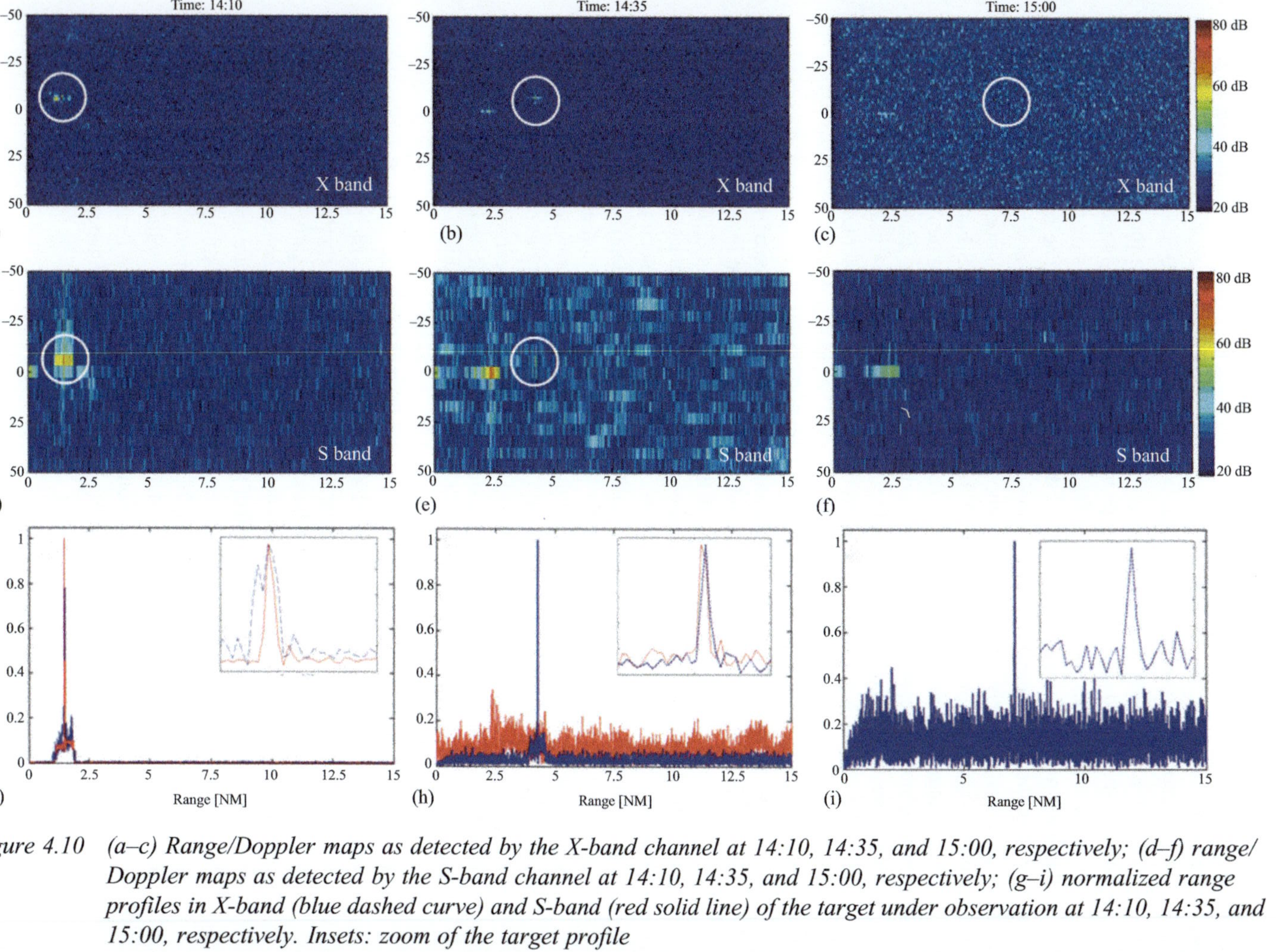

Figure 4.10 *(a–c) Range/Doppler maps as detected by the X-band channel at 14:10, 14:35, and 15:00, respectively; (d–f) range/Doppler maps as detected by the S-band channel at 14:10, 14:35, and 15:00, respectively; (g–i) normalized range profiles in X-band (blue dashed curve) and S-band (red solid line) of the target under observation at 14:10, 14:35, and 15:00, respectively. Insets: zoom of the target profile*

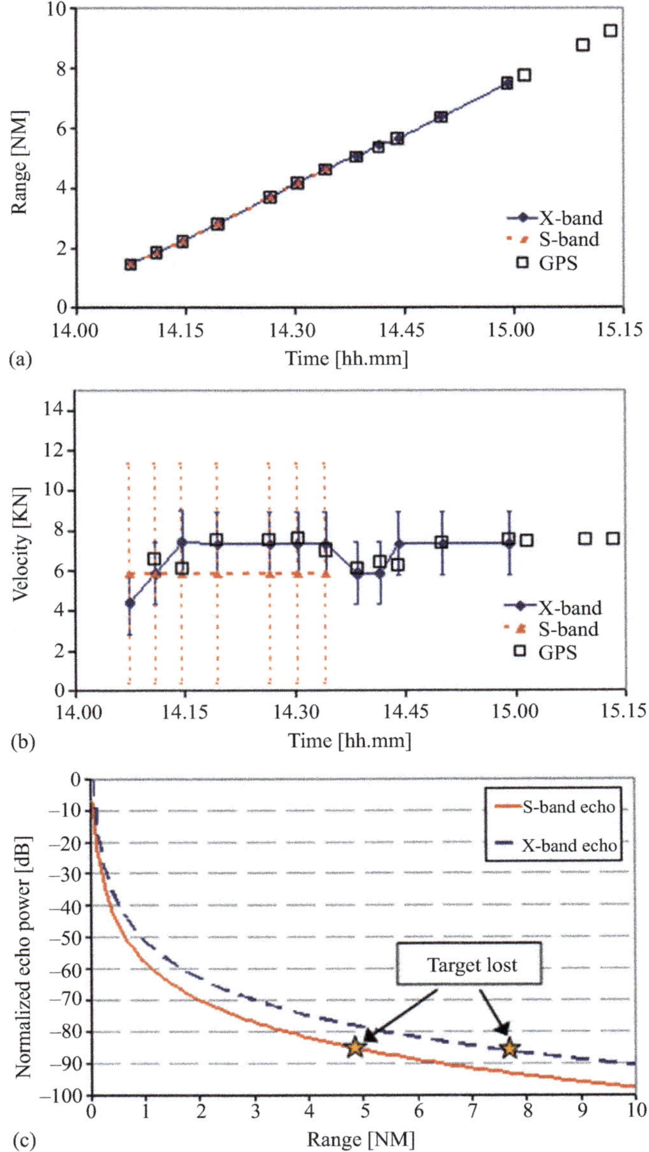

Figure 4.11 (a) Distance of the target from the radar site, as measured in the two frequency bands and provided by onboard GPS system; (b) velocity calculated from the Doppler shift in the two frequency bands and provided by the GPS; (c) estimated power of the echoes of the S-band and X-band signals at the radar receiver as a function of the target distance

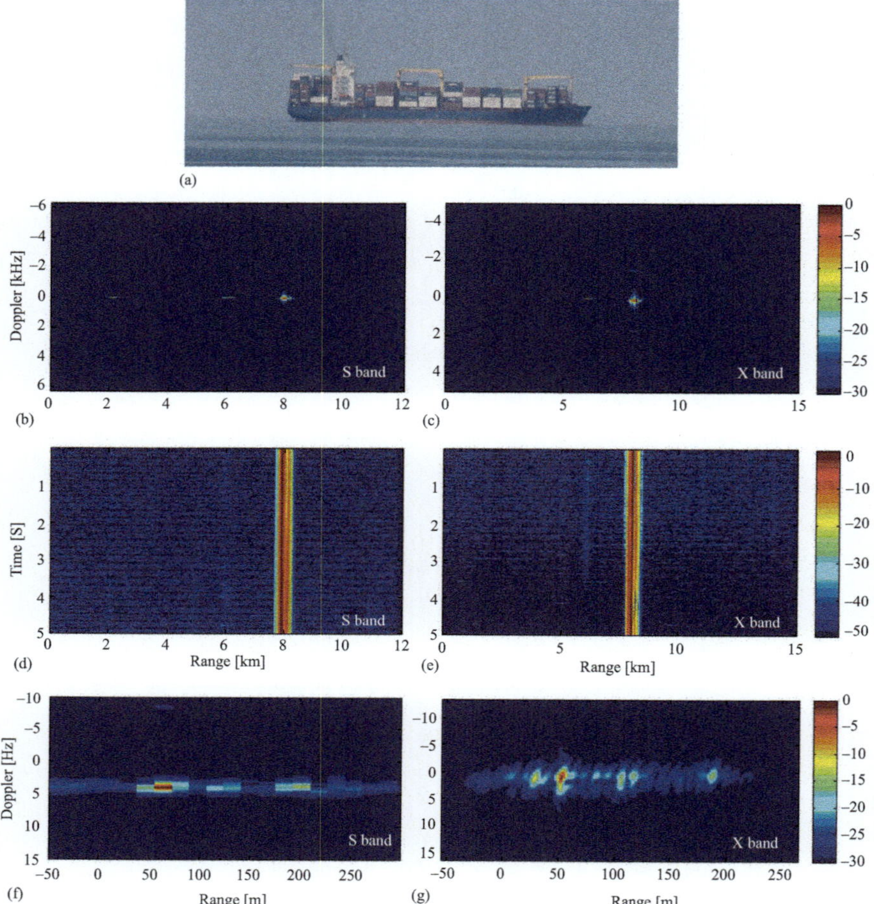

Figure 4.12 *(a) Picture of the detected target; (b) range Doppler map of the naval target as detected in S-band; (c) range Doppler map of the naval target as detected in X; (d) range profile of the moving target as detected in S-band; (e) range profile of the moving target as detected in X-band band; (f) ISAR image of the container ship as detected in S-band; (g) ISAR image of the container ship as detected in X-band*

configured as a linear chirp of 18 MHz over a pulsewidth of 2 μs, while the PRFs were set to 12.5 kHz for the S-band and 10 kHz for the X-band. The CIT was set to 20 ms, giving a Doppler frequency resolution of 50 Hz. Figure 4.12(b) and (c) report the radar detection of the moving target (besides other static vessels) in the S- and X-bands, showing respectively a Doppler component of about 50 Hz

(1st Doppler bin) and 150 Hz (3rd Doppler bin). These Doppler components correspond to a radial velocity of about 3 m/s and 2.3 m/s in the S- and X-bands. The increasing distance of the moving target is also visible in the range profiles of Figure 4.12(d) and (e). The calculated ISAR images in the S- and X-bands are reported in Figure 4.12(f) and (g). As can be seen comparing the ISAR images with the picture of the target (Figure 4.12(a)), the command station and the three crane are well represented.

4.9 Case study: aerial scenario field trial

The flexibility of the PANDORA system enabled by the use of photonics allows the dual band radar to adapt to largely different scenarios. In fact, the system described above has been used also in a field trial for aerial scenarios, as reported in [14], just by adapting the waveforms to objects moving at higher speed.

In more details, the measures were done by the group in Pisa from the roof of their lab, pointing at the air traffic from the close airport. The S- and X-band signals were both set to an 18 MHz linear frequency chirp over a PW of 2 μs, with PRF of 10 kHz. The CIT was set at 20 ms, thus integrating 200 radar pulses and obtaining a Doppler frequency resolution of 50 Hz. Figure 4.13(a) and (b) reports the range/velocity map of an aerial target exploiting the X- and S-band, respectively. The target is visible at a distance of about 4.6 km in both the maps and at a speed of about 55 m/s, which corresponds to a Doppler frequency of about 1.1 and 3.7 kHz for S- and X-band, respectively. As said above, the higher frequency of the X-band allows a higher velocity resolution. On the other hand, Figure 4.13(c) reports the range profile of the detected signals in the S-band (dashed red curve) and the X-band (dotted blue curve). As expected, their range resolution was almost the same and about 9 m since it depends on the signal bandwidth. On the other hand, as discussed in Section 4.5, the intrinsic coherence of the PANDORA system allows implementing an easy data fusion combining the information gathered in the two frequency bands. This was done with the data from the presented field trial, and the resulting range profile is shown in Figure 4.13(c) as well (solid black curve). The obtained range resolution is roughly halved with respect to a single frequency band, down to about 5 m.

In the aerial scenario as well, an ISAR imaging experiment has been run considering as the reference target a non-cooperating Boeing 737-800 with a total length and wingspan of about 40 and 35 m, respectively.

The radar detections in the two bands are depicted in Figure 4.14(a) (S-band) and (b) (X-band). The exploited ISAR processor was based on a windowed bidimensional Fourier transformation. First, an inverse FFT was performed on the complex samples to recover the high-resolution range profiles, and then an FFT was applied to each range bin, thus obtaining the cross-range dimensions. Figure 4.14(c) and (d) presents the extended range profile of the detected target as a function of the slow time (i.e., the sequence of detected radar pulses) in the S and X channels. A coherent processing interval of 2 s was selected in order to reduce

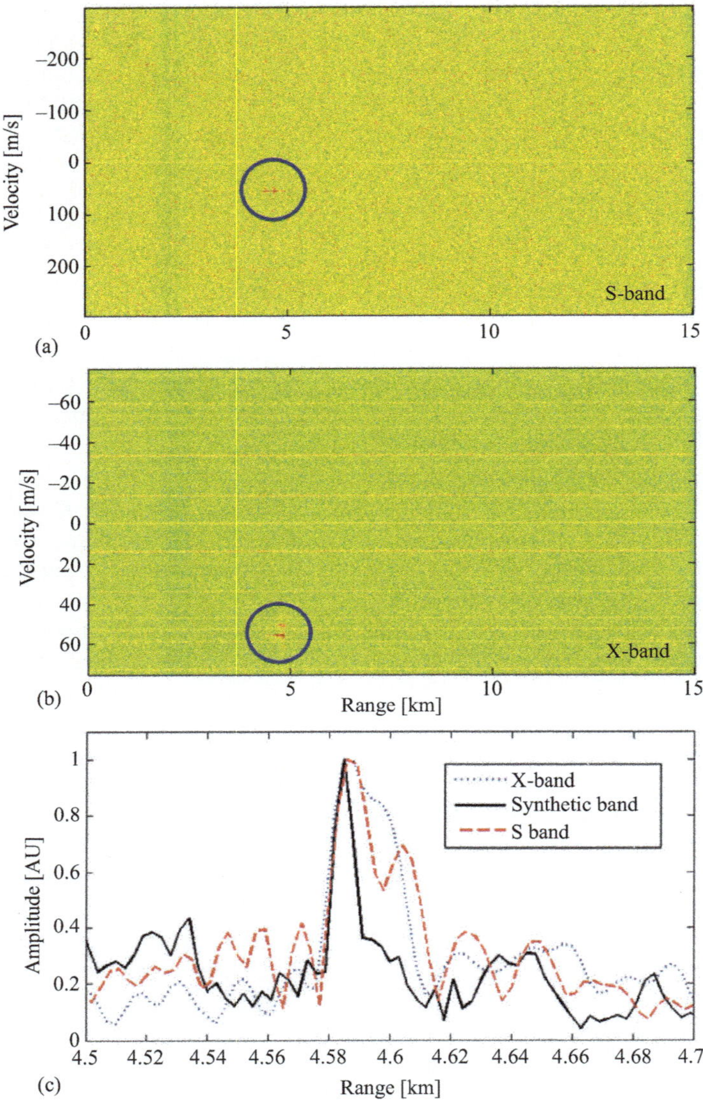

Figure 4.13 (a) Range velocity maps of the S-band and (b) X-band; (c) range profiles (C) of the dual band radar in an aerial scenario

the dimension of the data set. The ISAR images of the target (a Boeing 737-800) are reported in Figure 4.14(e) and (f). Although the image range and cross-range resolutions are coarse due to the limited resolution, a meaningful image of the target was obtained.

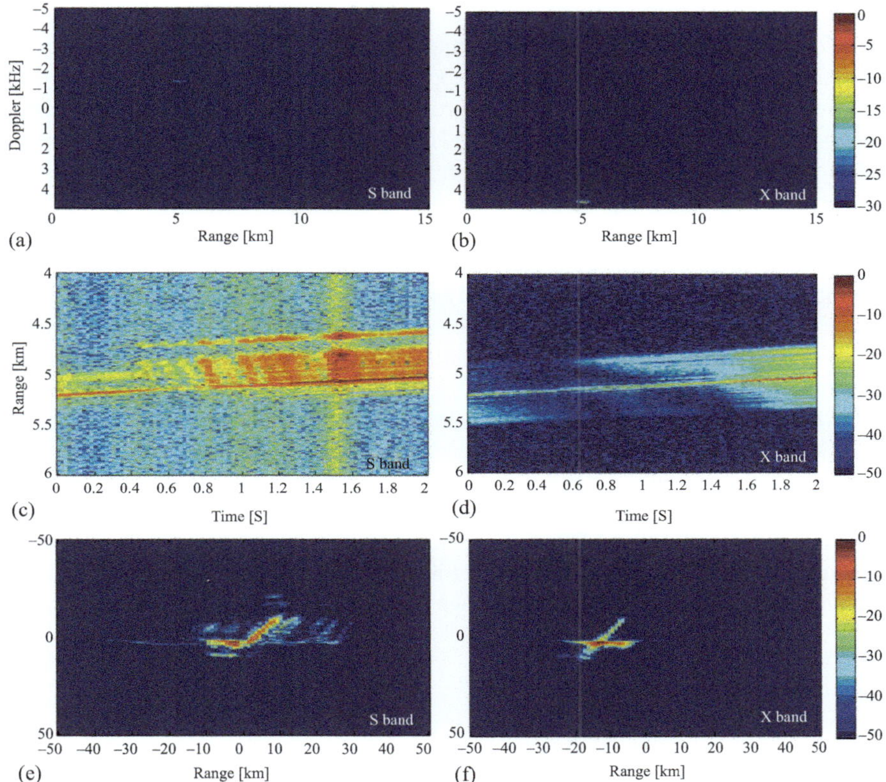

Figure 4.14 *(a) Range Doppler map of the aerial target as detected in S-band; (b) range Doppler map of the aerial target as detected in X; (c) range profile of the moving target as detected in S-band; (d) range profile of the moving target as detected in X-band; (e) ISAR image of the Boeing 737-800 as detected in S–band; (f) ISAR image of the Boeing 737-800 as detected in X-band*

4.10 Case study: environmental monitoring field trial

The flexibility enabled by the photonics-based architecture has also allowed another application: the environmental monitoring for ground displacement [15].

In the field of ground displacement measurements (e.g., for mine monitoring), radars with very high resolution (less than 1 mm) are necessary, and they also need to allow the displacement observation from very long distance (in the order of a few kilometers). The commercially available radar systems are based on differential interferometry implemented by means of the stepped-frequency continuous wave (SFCW) technique [16]. These kinds of radars change the frequency of the emitted CW signal in steps, with all the emitted sine waves coherent with each other.

The combination of several CW waveforms allows synthesizing a large signal bandwidth with minimal instantaneous bandwidth, thus improving the range resolution and reducing the noise. Moreover, applying differential phase algorithms, the range resolution can be significantly improved further. In these radars, therefore, the coherence between the generated CW frequencies is fundamental to ensure a correct phase estimation and thus a correct range shift estimation.

The commercial radar systems for environment monitoring all work in a single-frequency band. Nevertheless, a coherent multiband configuration would strongly benefit the displacement measure accuracy by applying the interferometric technique among the different (and largely detuned) frequency bands. Moreover, a flexible multiband system would allow tuning the operative RF carriers for adapting the system to the environment (as weather condition or observed scenario) and to the range of interest [17].

The researchers in [15] realized measurements of small displacements based on the SFCW technique and differential interferometry using a photonics-based dual-band coherent radar operating in S- and X-bands. In the work reported here, the target was illuminated by a sinusoidal signal that was stepped within frequencies f_n, coherent to each other and separated by Δf. Each frequency component of the backscattered echo accumulated a different phase shift, so that any change Δd in the distance induced a phase variation $\Delta\Phi$ between two frequencies separated by Δf [16], so that:

$$\Delta d = \frac{c}{4\pi \cdot \Delta f} \cdot \Delta\Phi \tag{4.1}$$

This way, the use of high Δf made it possible to detect displacements much smaller than a fraction of millimeter [18]. As an example, if one can detect a phase variation $\Delta\phi$ of 1° across 1 GHz, (4.1) sets the maximum resolution at 0.2 mm. Therefore, since the dual-band photonic transceiver guarantees the coherence between different RF bands, synthesizing a very large bandwidth with low-phase noise (thanks to the small noise bandwidth due to the narrow instantaneous bandwidth of the SFCW system), it allows a precise differential phase estimation that can be translated into very small displacement measures.

The scheme of principle of the multiband photonics-based radar system used to perform differential phase measurements is the same as reported in Figure 4.5. Here, the DDS was set to generate two different SFCW waveforms at two IFs (respectively $IF_S = 75$ MHz and $IF_X = 125$ MHz), so that they were up-converted to $CF_S = 2475$ MHz (S-band) and $CF_X = 9875$ MHz (X-band), as reported in Figure 4.15. Each sequence of stepped frequencies had 20 frequency steps spaced by 1 MHz with a duration of 200 μs each.

A wideband RF amplifier at the end of the transmitter chain boosted the RF signals to a horn antenna. A second horn antenna collected the RF echoes and the photonics-based receiver down-converted the signals back to IF and digitized it by a 400 MSample/s ADC. Here, the ADC also digitized a copy of the signal generated by the DDS in order to obtain the phase reference. The coherence was guaranteed by the MLL, which served as a reference for both up- and down-conversion of the radar signal.

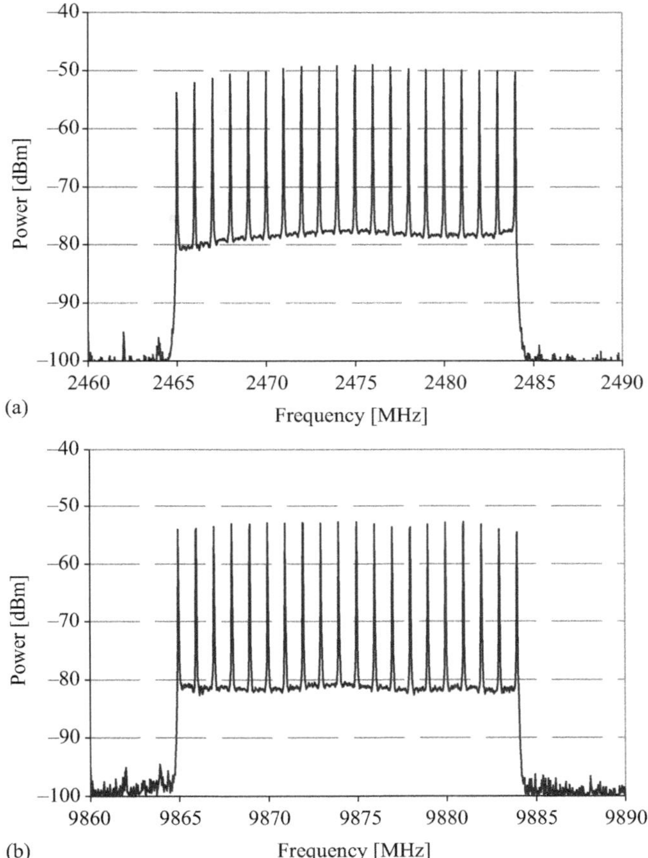

Figure 4.15 Electrical spectrum of the transmitted dual-band RF signal: (a) S-band signal; (b) X-band signal

Two broadband horn antennas with 60° beam width, one for signal transmission and one for the backscattered signal collection, were placed on a digitally controlled motorized linear platform (Figure 4.16). The displacements measurements were conveniently carried out by moving the antenna assembly, rather than the target.

The amplitude and phase spectrum of the received down-converted echo (Figure 4.17) and the reference signal (Figure 4.18) were first evaluated via standard FFT. The phase of the 40 SFCW tones was then extracted and the phase difference among the S- and X-band SFCW tones calculated in order to extract the actual target displacement by applying (4.1).

Considering the total bandwidth of 40 MHz covered by the SFCW signal, the range resolution of a standard radar would be 3.25 m. However, considering the dual-band operation, higher displacement resolution could be obtained since a

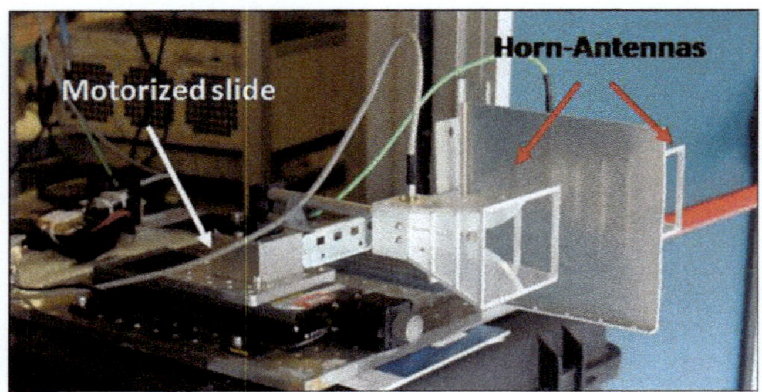

Figure 4.16 Moving antenna setup

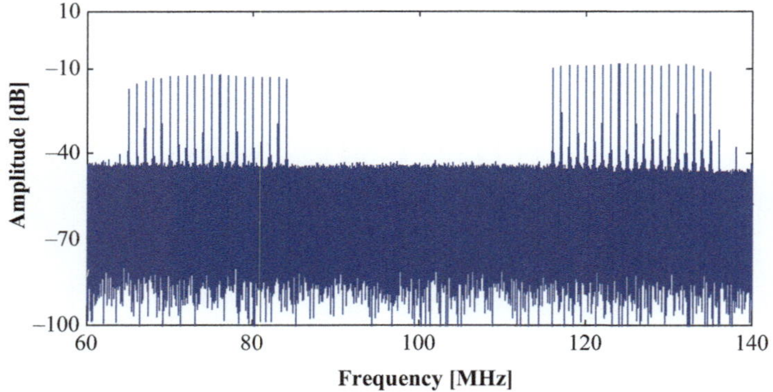

Figure 4.17 IF spectrum of the received SFCW signal

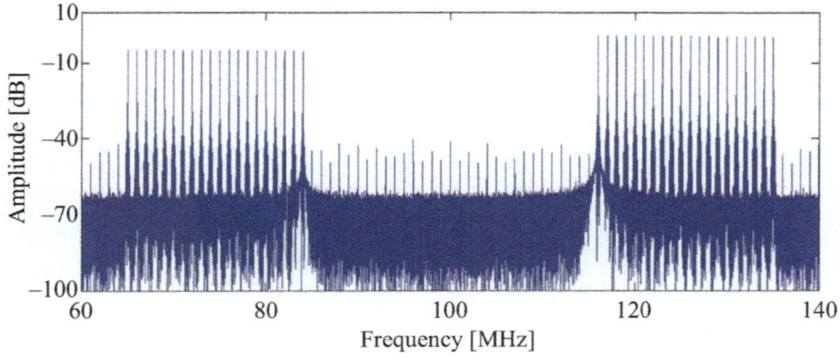

Figure 4.18 IF spectrum of the generated SFCW signal

maximum frequency difference of 7.4 GHz could be taken into account, giving a theoretical resolution of 20 mm. Moreover, considering the coherent analysis with the resolution expressed as in (4.1), we can expect a theoretical resolution 0.03 mm for one degree of phase detection.

In the experiment, the antenna was moved in steps as small as 0.2 mm, with the target (the laboratory wall) about 2 m from the radiating elements. For each step, a radar acquisition was made, the distance was calculated by means of the SFCW algorithm, and it was compared with the reference position (Figure 4.19(a)), giving a good agreement with the real target position, with a maximum measurement error limited to less than ±0.2 mm (Figure 4.19(b)).

As in a real application the target could be located up to few kilometers away from the radar, so the measures have been repeated emulating distances of 1.5 and 3 km by adding fiber spools. As can be seen in Figure 4.20, the error remained constant among the 3 km target distance and below 0.2 mm, thus confirming the phase stability of the photonics-based radar system.

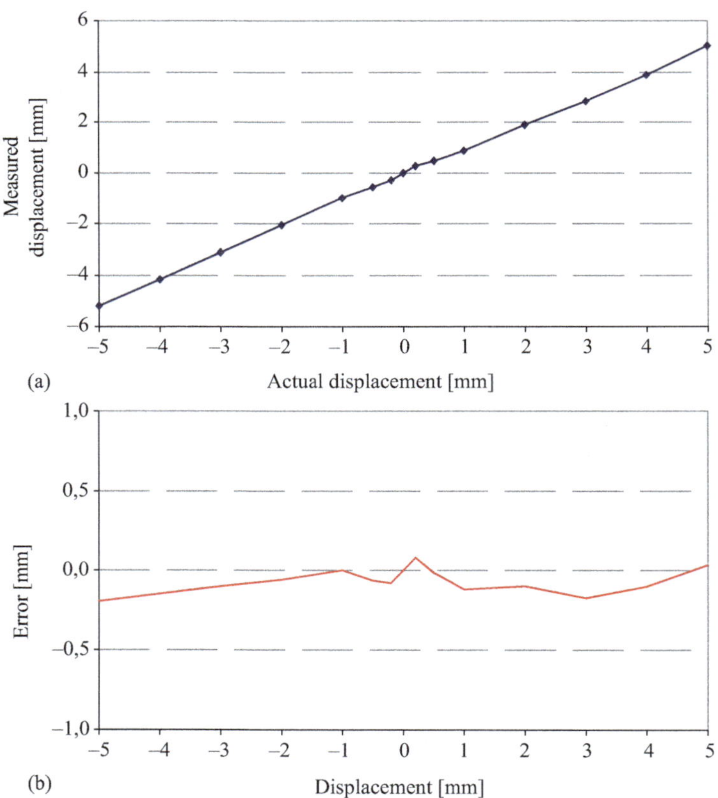

Figure 4.19 (a) Measured displacement versus real displacement; (b) displacement measure error

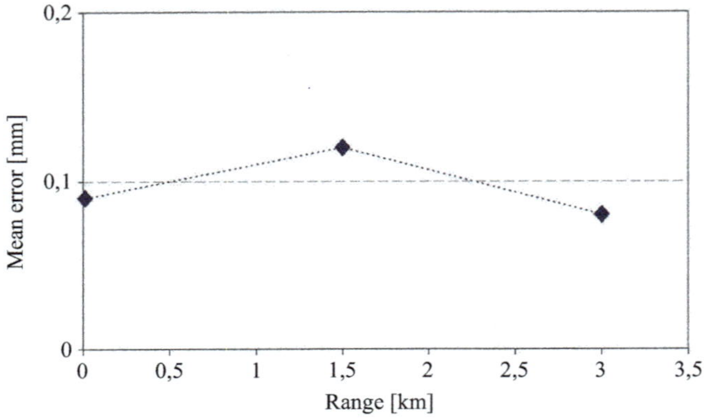

Figure 4.20 A mean error of the displacement measurement error

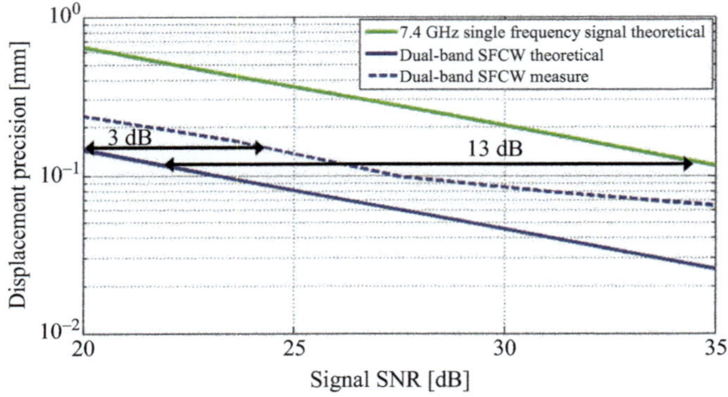

Figure 4.21 Displacement precision versus signal SNR

An important limiting factor in the phase estimation is the SNR of the detected signal. In Figure 4.21, the precision of the measured displacement, expressed in millimeter, is reported as a function of the signal SNR. The green curve shows the theoretical displacement precision for a single 7.4 GHz sinusoidal signal (equal to the frequency difference among the S- and X-bands). Such curve represents the maximum accuracy achievable by the system if transmitting a single sinusoidal tone for each of the two bands rather than the SFCW Instead, the blue lines show the displacement precision achieved by the SFCW signal with 20 frequency steps in each frequency band, in the ideal case (blue solid line, simulated result) and in the measured one (dotted line). The SFCW curve experienced 13 dB gain compared to the single frequency case, which can be ascribed to the SNR improvement given by the averaging gain across 20 independent phase measurements.

With respect to the simulated displacement precision, the measured precision reports an SNR penalty of 3 dB for low SNR, increasing at higher SNR, probably due to system saturations (e.g., amplifiers, ADCs, etc.).

4.11 Summary

In this chapter, we have reported the most significant examples of photonics-based radar systems, analyzing the transceivers structures, the processing architectures, and the filed trial case studies.

Concerning the single-band radar approaches, it is particularly important to underline the reported comparison between a photonics-based radar prototype and a commercial radar system for maritime applications. In fact, the reported performance is very similar although the photonics-based radar is still at a prototype stage. This demonstrates that photonics can already reach the same results of standard RF systems. Moreover, the flexibility and the precision of photonics clearly enables new possibilities, paving the way for the software-defined radars. A second example of photonics-based radar has been also reported, demonstrating an impressively ultra-wideband radar with as much as 8 GHz of instantaneous bandwidth, ensuring a correspondingly fine range resolution, while its receiver architecture can be managed by low-speed ADCs so that an ISAR processing can be implemented in real time.

On the other hand, several paragraphs are devoted to the description of the coherent multiband radar approach, which is a peculiar feature of photonics-based radars. As a confirmation of the flexibility brought by using photonics, the reported case studies deal with very different scenarios: maritime, aerial, and ground displacement: the software-defined approach enable by photonics has allowed to exploit the very same architecture for each of these applications, with results aligned with state-of-the-art systems. In particular, in the case of sub-millimeter displacement, the availability of two largely detuned coherent RF bands has permitted highly precise displacement measures.

In conclusion, the reported results clearly draw the advancement state of the photonics-based radar, highlighting the strong innovation potentials of the photonic approach.

References

[1] P. Ghelfi, F. Laghezza, F. Scotti, *et al.*, “A fully photonics-based coherent radar system”, Nature, vol. 507, no. 7492, pp. 341–345, 2014.

[2] F. Zhang, Q. Guo, Z. Wang, *et al.*, “Photonics-based broadband radar for high resolution and real-time inverse synthetic aperture imaging”, Opt. Exp., vol. 25, no. 14, pp. 16274–16281, 2017.

[3] T. Nagatsuma, S. Horiguchi, Y. Minamikata, *et al.*, “Terahertz wireless communications based on photonics technologies”, Opt. Exp., vol. 21, no. 20, pp. 23736–23747, 2013.

[4] F. Laghezza, F. Scotti, P. Ghelfi, and A. Bogoni, "Photonics-assisted multiband RF transceiver for wireless communications", IEEE J. Lightwave Technol., vol. 32, no. 16, pp. 2896–2904, 2014.

[5] F. Scotti, F. Laghezza, G. Serafino, *et al.*, "In-field experiments of the first photonics-based software-defined coherent radar", J. Lightwave Technol., vol. 32, no. 20, pp. 3365–3372, 2014.

[6] F. Laghezza, F. Scotti, G. Serafino, *et al.*, "Field evaluation of a photonics-based radar system in a maritime environment compared to a reference commercial sensor", IET Radar Sonar Navig., IET Radar Sonar Navig., vol. 9, no. 8, pp. 1040–1046, 2015.

[7] F. Scotti, F. Laghezza, P. Ghelfi, and A. Bogoni, "Multi-band software-defined coherent radar based on a single photonic transceiver", IEEE Trans. Microwave Theory Techn., vol. 63, no. 2, pp. 546–552, 2015.

[8] F Scotti, F Laghezza, D Onori, and A Bogoni, "Field trial of a photonics-based dual-band fully coherent radar system in a maritime scenario", IET Radar Sonar Navig. vol. 11, no. 3, 2016.

[9] P. Ghelfi, F. Laghezza, F. Scotti, D. Onori, and A. Bogoni, "Photonics for radars operating on multiple coherent bands", invited paper, J. Lightwave Technol., vol. 33, no. 2, pp. 500–507, 2016.

[10] M. Vespe, C. J. Baker, and H. D. Griffiths, "Automatic target recognition using multi-diversity radar," IET Radar Sonar Navig., vol. 1, no. 6, pp. 470–478, 2007.

[11] P. Van Dorp, R. Ebeling, and A. G. Huizing, "High resolution radar imaging using coherent multiband processing techniques," in Proc. IEEE Radar Conf., Washington, DC, USA, 2010, pp. 981–986.

[12] X. Wei, Y. Zheng, Z. Cui, and Q. Wang, "Multi-band SAR images fusion using the EM algorithm in contourlet domain," in Proc. 4th Int. Conf. Fuzzy Syst. Knowl. Discovery, Haikou, China, pp. 502–506, 2007.

[13] "IALA VTS Manual", www.iala-aism.org/products/publications, accessed April 2016.

[14] F. Laghezza, F. Scotti, D. Onori, and A. Bogoni, "ISAR imaging of non-cooperative targets via dual band photonics-based radar system", International Radar Symposium (IRS), Krakow, Poland, 2016.

[15] S. Pinna, S. Melo, F. Laghezza, et al., "Photonics-based radar for sub-mm displacement sensing", IEEE J. Sel. T. Quantum. Elec., vol. 23, no. 2, 5300408, 2017.

[16] I. Nicolaescu, P. van Genderen, K. W. Van Dongen, and J. Van Heijenoort, "Stepped frequency continuous wave radar-data preprocessing", in Proc. Adv. Gr. Penetrat. Radar, pp. 177–182, Delft, Netherlands, 2003.

[17] J. D. Taylor, and P. E. Taylor. "Ultra-wideband radar technology." CRC, London (2001).

[18] K. Meiners-Hagen, R. Schödel, F. Pollinger, and A. Abou-Zeid, "Multi-wavelength interferometry for length measurements using diode lasers", Measur. Sci. Rev., vol. 9, no. 1, 2009.

Chapter 5
Radar networks

Carlo Noviello[1], Paolo Braca[2] and Salvatore Maresca[3]

5.1 Chapter organization and key points

With the advent of innovative electronic technologies and new digital signal processing techniques, radar systems have greatly enhanced their performance in terms of detection, tracking, and imaging capabilities. When a target is illuminated by electromagnetic radiation, the scattered signal is reflected in all directions and, therefore, a single sensor can only intercept a small part of the overall reflected energy. In this way, most of the information is lost. However, the majority of current radar systems still operate in monostatic configuration (i.e., with only one antenna for both signal transmission and reception). On the other hand, radar networks have the potential to increase the capabilities and performance of current radar systems by exploiting their intrinsic multiperspective observation capability.

The aim of this chapter is, thus, not only to underline the advantages of radar networks, but also to describe the major technical issues for their correct operation. First, a brief introduction to the concept of radar networks will be provided, together with an overview of the main benefits and the most relevant applications. Secondly, the problem of radar network categorization has been examined. A first categorization is based on the different transmitting and receiving options of the network sensors, which can be monostatic, bistatic (i.e., the transmitting and receiving antennas are widely separated) or a combination of mono/bistatic. A second categorization is based on the distinction between centralized and decentralized processing approaches instead. In particular, the advantages and disadvantages of centralized and decentralized approaches will be considered, with particular attention to the final system performance and additional complexity trade-offs. Third, the chapter will extensively discuss the network synchronization and data fusion issues, which are vital to ensure excellent operation of the radar network. Synchronization directly impacts on radar network performance. For this reason, the main causes that could lead to the mis-synchronization of the radar

[1]Institute for Electromagnetic Sensing of the Environment, Naples, Italy
[2]Maritime Research and Experimentation, Pisa, Italy
[3]Institute of Technologies for Communication, Information and Perception (TeCIP) – Sant'Anna School of Advanced Studies, Pisa, Italy

network elements will be presented. Later, the data fusion problem in the context of the radar networks will be investigated with particular emphasis on the multitarget tracking (MTT) issue.

Finally, the chapter concludes with a recent scientific experimentation conducted by the NATO Science and Technology Organization Centre for Maritime Research and Experimentation, in which a network composed by two high-frequency surface wave (HFSW) radar systems was deployed in the Ligurian and northern Tyrrhenian Seas for maritime surveillance. The description of the experimental example is aimed to prove the effectiveness and the scientific relevance of radar networks and their decisive impact for homeland security and human progress.

5.2 Multistatic radars

In the last decades, the majority of radar systems have been developed as single monostatic sensors, essentially due to their simplicity and the wide range of applications they can be exploited for. However, for a monostatic radar, sensitivity is limited by the product of the transmitted power and antenna aperture, while location accuracy is limited by the direct radar-target perspective. The majority of these limitations can now be handled by adopting multiple transmitters and receivers, thus leading to the concept of multistatic radar systems [1]. However, before going deep into details, it is important to understand that the definition of multistatic radar is not univocally consistent within the current literature. In general, multistatic radars monitor areas that are covered by multiple transmitters and receivers. The information coming from the multiple sensors is combined together (i.e., fused) to provide improved target detection, tracking, and imaging capabilities. This allows to differentiate multistatic radar systems from radar networks, which monitor independently different spatial coverage regions and share the information together in order to increase the coverage area. However, the terms radar network and multistatic radars are frequently adopted to describe the same form of system in the literature [2]. Moreover, the so-called multiple-input multiple-output (MIMO) radar systems have recently gained significant attention. We can consider them as a subset of multistatic radar systems; however, again there are no rigorous definitions for discriminating MIMO radars [3] from the concepts of multistatic radars and radar networks [4].

5.2.1 Concept

Nowadays, radar systems are essential monitoring sensors in a wide range of different civil and military contexts and in many remote sensing applications. The main reason can be found in their ability to monitor very large areas in any weather condition and illumination (e.g., clouds, fog, ash, powder, etc.). Many countries have a network of radars for controlling the air traffic, to monitor the maritime coasts and the terrestrial activities. Radar networks are thus an essential part of a wider defense capability. In fact, radar networks are able to detect aircrafts, ships, and any other possible moving object that can be a possible hostile threat. Now, however, targets are becoming faster, very agile, and stealthier; therefore, the nature of hostile threats is becoming uncontrolled and almost unlimited [5]. In such scenarios, unfortunately, conventional radar systems could not grant the requested level of reliability. For this reason, the scientific community is focusing its attention on multistatic radar and radar network concepts [6].

Currently, radar networks are of emerging interest as they can exploit spatial diversity, enabling improved performance and developing new kinds of applications. As pointed out, such systems differ from typical active radar systems; in fact, they consist of multiple spatially diverse transmitter and receiver sensors that may potentially cooperate together [7]. However, radar networks are very complex systems and this relevant complexity implies many technical issues. The most important challenges are essentially timing and frequency synchronization and the data fusion of the information from the radar network elements [8].

Time and frequency synchronization is very crucial for the correct operation of radar networks. Often, due to terrain roughness, presence of lakes, sea, or due to time limitations, establishing a common time and frequency reference by means of cable or fiber could be difficult or even impossible. Alternatively, a wireless communication link could be used, but the presence of multipath would degrade the time accuracy. To overcome the aforementioned limitations, currently, the most important technology that has been adopted to synchronize radar networks is the Global Positioning System (GPS). The GPS signal is exploited as a reference timing signal for all the radar network nodes (i.e., each transmit/receive pair in the network) [9].

Sensor data fusion is also very important since the distributed nature of the radar sensors requires a distributed approach to effectively combine the data information stream. Obviously, the data exchange process implies the necessity of reliable, effective, and high bandwidth communication links among the network nodes [10].

5.2.2 Benefits

In general, radar networks have a number of advantages that make them potentially very well suited for many applications if compared with stand-alone operating radars. In fact, multistatic radar sensors do not necessarily require complicated and expensive transmitters. Thanks to the distributed locations of the transmitters and the inherently passive nature of the multiple receivers, radar networks are also relatively immune to electronic and physical attacks. They are often able to detect stealthy targets, which are specifically designed to exhibit a very small radar cross-section (RCS) against monostatic radars [11]. In fact, they grant increased sensitivity by collecting the scattered energy coming from multiple directions, and hence improving detection and tracking performance. Furthermore, target imaging, classification, and recognition can also be enhanced as the target can be observed from different angular perspectives [12]. However, these advantages come at the cost of increased hardware and software workload. Nevertheless, the necessary trade-off between performance and increased global system complexity may still provide worthwhile benefits. In fact, the separation among transmitters and receivers in a multistatic radar offers the system engineers new and additional degrees of freedom to devise innovative solutions for many specific applications. Moreover, increased survivability and reliability can be achieved thanks to the option of having passive receivers. These receivers can improve the discovery accuracy of possible malicious jammers by sharing and fusing the information from the distributed network nodes [8]. Furthermore, the dislocation of multiple transmitters and receivers in different zones, on one hand, can improve the coverage area, and, more importantly, by observing the target from different line of sights, they also improve the detections of stealthy targets. These advantages make multistatic radar systems very attractive for a number of applications, many of which could fit well with the needs of homeland security and electronic warfare [13]. At the same time, however, the further complexity of tackling an increased number of distributed transmitters and receivers brings about new scientific challenges that require careful understanding if these kinds of sensors should be adopted for operational and real-time usage.

5.2.3 Applications

Currently, many scientists claim that the most recent technology developments such as parallel computing, GPS, wideband wireless communications, and array antennas are a sort of precursors for the upcoming development of innovative operational radar network systems. Radar networks are widely adopted for both military and civilian applications. For military defense purposes, multistatic radars can be used to form a tailored surveillance area in order to efficiently detect non-cooperative targets. In fact, many are the degrees of freedom that allow to improve system performance under different operational conditions. Among these we can mention multistatic geometry, different antennas, multiple signal waveforms, carrier frequencies (operational frequency bands), and polarization for each node of the radar network. These parameters may be modified according to the specific applications and the context of interest, either for air, maritime, and terrestrial defense activities [1]. The same concept can be also used for underwater surveillance by the use of sonar networks [14–16]. In addition, as we previously mentioned, detection of stealth targets can be significantly improved by adopting multistatic radar systems. In fact, stealth objects are designed to be undetectable only against monostatic radar systems. In part, this property can be achieved by allowing the signal to be scattered in different directions with respect to monostatic line of sight. In addition, stealth targets adopt signal absorption techniques for further decreasing the backscattered RCS. This does not apply at bistatic or multistatic angles as the aircraft's backside. Obviously, one simple solution is to place the receiver far away from the transmitter (i.e., bistatic setup). In this way, the system will measure the bistatic RCS of the target. An alternative possibility, however more expensive respect to the previous one, is the use of a second monostatic radar in the opposite location with respect to the first one. Thus, when the first radar is in the blind zone of the stealth target, the second radar will be able to detect, track, and construct an image of the target.

A number of civilian applications are also worth to be mentioned. For example, common Air Traffic Control systems benefit from the use of multistatic radars rather than single monostatic systems, counteracting the electromagnetic multipath effect, which degrades the performance. In this way, it is possible to enhance the civilian safety by monitoring more appropriately the surveillance area [17]. A multistatic Ground Penetrating Radar system is also presented in [18] for the detection of antipersonnel mines. Radar networks can be also utilized to detect and classify airplanes, helicopters, and boats. By exploiting multiple observation perspectives, it is possible to discern the different components of the target, such as wings, rotors, and helixes. Moreover, sensor networks are also widely adopted in the automotive sector for vehicular management and safety purposes. In this context, many are the important vehicle system parameters that can be extracted, such as the velocity, the pitch angle, the distance w.r.t. the vehicle and the ground or between two vehicles, and, finally, also the road surface conditions [19]. All these parameters are very critical for the development and correct operation of reliable braking and collision avoidance systems, lane-change assist, vehicle blind-spot detection, airbags arming, and pre-crash early warning systems, etc.

5.2.4 Network radar: system description

In this section, a closer description of radar network architectures, signal models, and the radar network processing approaches is investigated. In fact, in the current literature, radar network architectures can be categorized by considering several network aspects: the geometry, the processing methods, the sensors specification, the complexity, etc. [8]. In particular, in this section, we propose a first categorization approach based on the different transmitting and receiving options of the sensor network, and, then, a second categorization based on signal processing is examined.

Concerning the sensor combination options, one of the principal aspects for differentiating radar networks is related to the transmitting and receiving management of the network elements. In fact, the sensors of the network may be grouped into three main categories: (1) monostatic, (2) bistatic, and (3) a combination of mono/bistatic (multistatic) radar systems.

In the monostatic case, each radar system of the network transmits a specific signal and each sensor receives only the echo that has been originated by itself. An example of this case is shown in Figure 5.1. This is the simplest way to build up a radar network as there is no necessity to synchronize the networks elements, and, therefore, the overall system processing is very simplified.

In the second case, the network is composed by bistatic radars, as depicted in Figure 5.2. Here, the radar network is composed of one common emitter and N spatially separated receivers. As evident from Figure 5.2, the transmitter in combination with each single receiver forms a transmitter–receiver pair that can be identified as single bistatic radar system. This bistatic network system can exploit

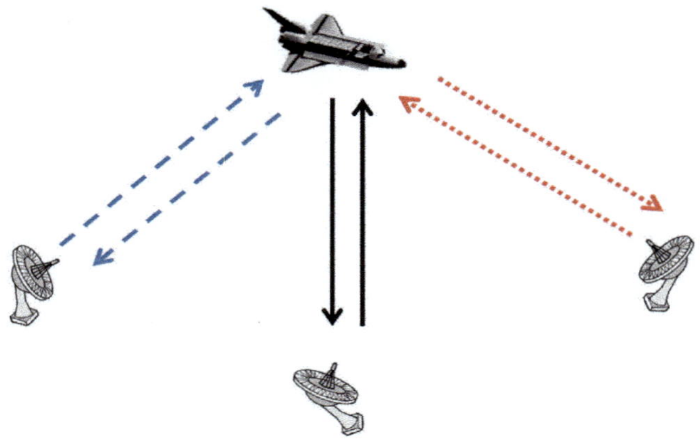

Figure 5.1 Monostatic radar network system

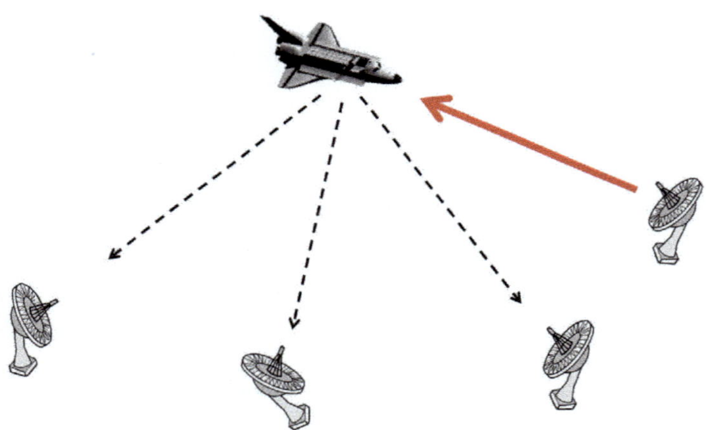

Figure 5.2 Bistatic network system

the advantages of classic bistatic radars as the robustness against jammers, the simplicity of the multiple receiving systems, and the capability of detecting stealth targets. However, these benefits are paid in terms of higher complexity w.r.t. the case a network is composed by only monostatic radars.

The combination of monostatic and bistatic can be considered as the most general case. In fact, each transmitter–receiver pair of the radar network, typically called node, receives its own and the signals backscattered from the targets, which are transmitted by the other nodes (see Figure 5.3). Transmitter–receiver path pairs are typically indicated as nodes of the network. As described in Figure 5.3, by assuming three transmitted–receiver pairs, the radar network may be considered

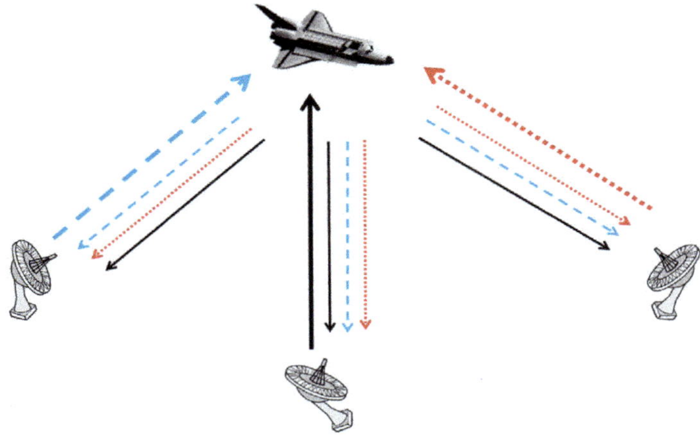

Figure 5.3 Multistatic radar network system

composed of nine nodes, which corresponds to the number of signals to be processed. The multistatic network case combines the advantages of monostatic and bistatic network typologies, thus achieving flexibility and robustness at the expense of an increased system complexity. In fact, to increase system sensitivity, the waveforms transmitted by each sensor must be orthogonal to each other. Moreover, due to the angular diversity, all the signals received by each sensor can be assumed spatially independent because each antenna observes the target from a different perspective. In fact, each antenna observes the target from a different perspective and so the RCS measured by each receiver can deeply differ from other measurements acquired by the other sensor. Thanks to the aforementioned features, assuming a robust synchronization among the nodes of the network, and adequately processing the radar network data, the following benefits can be obtained: (1) robustness against target scintillation, (2) improved detection capability, (3) improved target position estimation, (4) capability of resolving multiple closely spaced targets, and (5) increased contrast against jammers and, hence, strong electronic counter-countermeasures ability.

A different way to categorize radar networks is considering the alternative ways of processing the data acquired by the radar nodes. There are two main processing options that group radar networks essentially into centralized and decentralized systems. As previously indicated, there are many parameters to be considered [20]: the sensitivity and the robustness of the network. The system sensitivity depends on the communication and processing capabilities of the network. The larger is the volume of the data that can be shared among the sensor network elements, the better is the overall sensitivity. To obtain very high system sensitivity, it is necessary to transmit huge volumes of data. The centralized processing approach tries to maximize the sensitivity of the overall system by adopting a central processing unit, which firstly collects all the raw data acquired by the

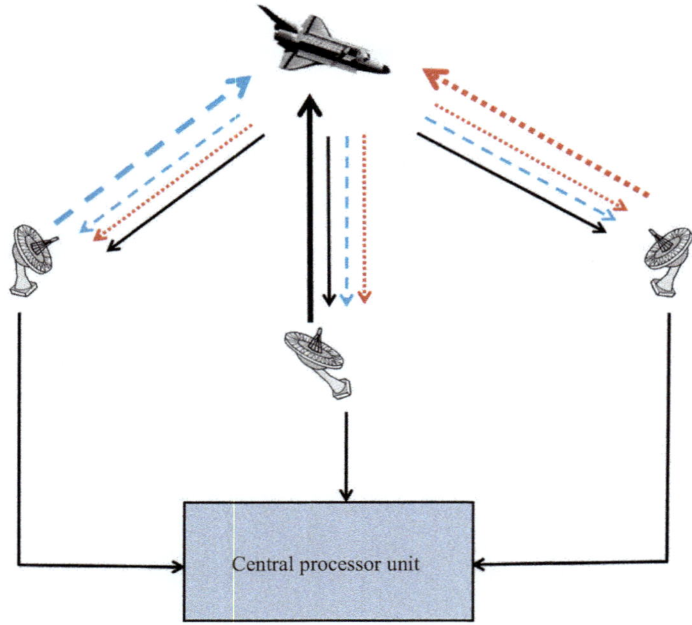

Figure 5.4 Centralized processing approach

sensors (without any preprocessing step) and then combines all the data together for the final decision (see Figure 5.4). However, this approach requires very large communication bandwidth and may result to be very expensive.

Alternatively, in the decentralized processing approach, each radar can perform some data preprocessing before transmitting them to the central coordinator station. The preprocessing results are then sent to the central processing unit, where the final decision-making is carried out. The decentralized processing is shown in Figure 5.5 [8].

To better understand the relationship between the centralized and decentralized processing, let us introduce the netted radar network signal model.

5.3 Signal model

In the most general case, a radar network can be assumed to be composed by N transmitter modules (Tx) and M receivers (Rx), which are spatially separated. Each receiver contains N matched filters for each of the N transmitted signals. Therefore, a total of $N \times M$ signals are available for the processor unit. Transmitters and receivers, which form the sensing elements of the network, have typically the same system specification (the same carrier frequency, the same bandwidth, the same power, antenna, etc.). Moreover, they are geographically separated to ensure the aforementioned multiperspective by exploiting the angular diversity. Let us introduce

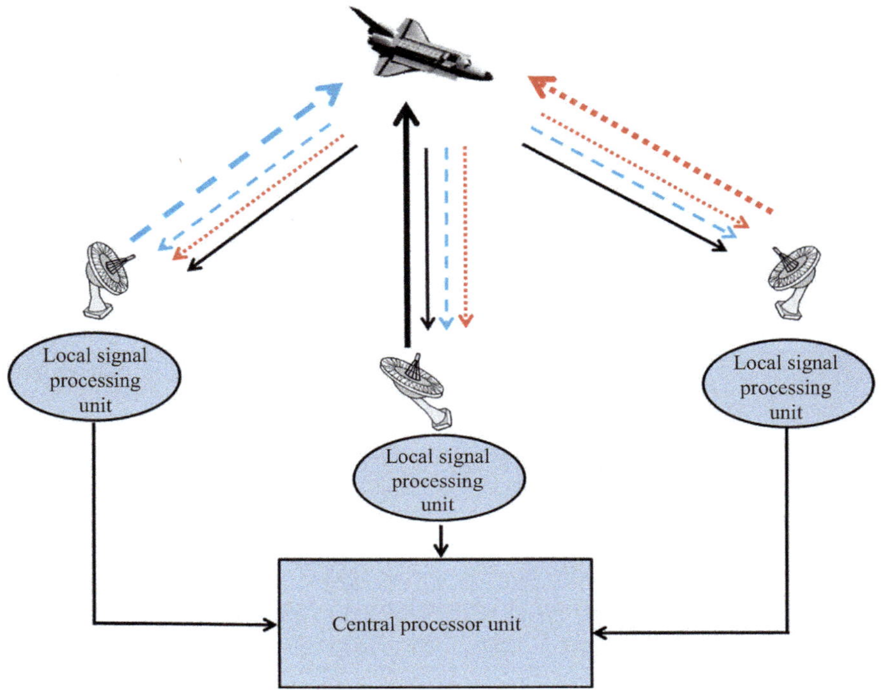

Figure 5.5 Decentralized processing approach

a simple system signal model. The signal arriving at the mth receiver can be described as follows [21]:

$$r_m(t) = \sum_{n=1}^{N} \left[H_{\substack{0\\1}} \alpha_{n,m}(\sigma) s_n(t - \tau_{n,m}) + c_{n,m}(t - T_{n,m}) \right] + I_m(t) + z_m(t), \quad (5.1)$$

where t is the time variable, $H_{0/1}$ is 0 or 1 depending on the presence or the absence of the target, respectively; s_n is the nth transmitted signal; $c_{n,m}$ represents the clutter contribution; z_m is the electronic thermal noise; I_m is the external disturbance (e.g., jamming); $\tau_{n,m}$ is the delay that accumulates during the path between the nth transmitter and the target and respectively from the target to the mth receiver; and $T_{n,m}$ is the delay between the nth transmitter and the clutter cell for the mth receiver. Finally, $\alpha_{n,m}(\sigma)$ is a coefficient that accounts for the signal phase, the RCS distribution, and the radar equation, as follows [21]:

$$\alpha_{m,n} = \sqrt{\frac{P_{Tx}}{N}} \sqrt{\frac{G_{Tx} G_{Rx} \lambda^2 \sigma}{(4\pi)^3 R_n^2 R_m^2}} exp\left\{ -j \frac{2\pi R_{n,m}}{\lambda} \right\}, \quad (5.2)$$

where G_{Tx} and G_{Rx} are respectively the gains of the transmitting and receiving antennas; σ is the RCS of the target; P_{Tx} is the transmitted power; R_n and R_m are the transmitter–target and target–receiver distances, respectively, and $R_{n,m}$ is the round-trip distance traveled by the electromagnetic signal [21]. By neglecting the presence of clutter, the received signal, due to the matching filtering operation, may be expressed as the result of the cross-correlation between the received signal and the system impulsive response, as follows [21]:

$$x_{h,m} = H_{0/1}\alpha_{h,m}(\sigma)R_h(k - \tau_{h,m}) + H_{0/1}\sum_{n=1,n\neq h}^{N} \alpha_{n,m}(\sigma)R_{n,h}(k - \tau_{n,m}) + n_{h,m}(m) \tag{5.3}$$

where R_h is the autocorrelation function (relative to the matching function $h(\cdot)$) of s_n and $R_{n,h}$is the cross-correlation function between s_n and s_h, and $n_{h,m}$ is the component of the overall disturbance collected by the mth receiver after the matched filter operation. The last two terms in (5.3) represent the overall contribution of the thermal noise at the receiver front-end. This equation highlights how the use of orthogonal waveforms is essential to separate and discern the transmitted waveforms from each other. For the sake of simplicity, let us assume an arbitrary radar network geometry for which two different types of system processing solutions are going to be analyzed. Indeed, the only difference between the two systems is the way the received signals are processed. Moreover, by assuming that network synchronization is granted, all the signals can be written in matricial form as follows [21]:

$$\boldsymbol{A} = \begin{pmatrix} \chi_{11} & \chi_{12} & \cdots & \chi_{1N} \\ \chi_{21} & \chi_{22} & \cdots & \chi_{2N} \\ \vdots & \vdots & \vdots & \vdots \\ \chi_{M1} & \chi_{M2} & \cdots & \chi_{MN} \end{pmatrix} \tag{5.4}$$

Alternatively, the previous formula can be rewritten by rearranging the matrix elements into a single vector $\boldsymbol{\chi}$ as follows:

$$\boldsymbol{\chi} = \left[\chi_{11}, \chi_{12} \cdots, \chi_{1N}, \chi_{21}, \chi_{22}, \cdots, \chi_{2N}, \cdots, \chi_{M1}, \chi_{M2}, \cdots, \chi_{MN}\right]^T. \tag{5.5}$$

5.3.1 Centralized radar network processing

In the centralized radar network approach, after the matching filtering process, the signal is coherently integrated without any preprocessing stage. The phases of the received signals are highly correlated as they deeply depend on the target location and on the acquisition geometric system. However, the phases of signals are typically described by uniform random samples, distributed between $[-\pi, \pi]$; in fact, the phase of the signal is wrapped every half-wavelength; therefore, the target's position cannot be measured with the same precision of the radar wavelength accuracy. In this centralized approach, if all the phases are uniformly distributed, then the coherent phase integration will correspond to a signal whose amplitude may be less than the sum of the amplitudes recorded at each radar network element. In the worst case, when the amplitude is constant and the result of the phase summation is 2π, the integrated signal can be completely cancelled out. In such extremely condition, this processing strategy will provide the lower bound performance limit, and, the resulting signal-to-noise ratio (SNR), after the integration operation, will be comparable to the single sensor case. In the centralized radar network processing, the samples are summed coherently and the relative signal power is then compared with the threshold for the detection stage. In this case, the coherent detector can be written as:

$$\left|\sum_{k=1}^{N} \chi_k\right|^2 \underset{H_0}{\overset{H_1}{\lesseqgtr}} \eta, \tag{5.6}$$

where η is a suitable decision threshold. Another coherent radar network processing is the rephased coherent processing. The rephased approach adopts the same samples as the previous cases, but the central processor unit perform an additional step. The aim of the additional step is to align the phases of the vector signal χ defined in formula (5.5) according to the exact position of the target, thus maximizing the SNR and subsequently improving the global system detection performance as follows [21]:

$$\left|\sum_{k=1}^{NL} \chi_k e^{-j\vartheta_k}\right|^2 \underset{H_0}{\overset{H_1}{\lesseqgtr}} \eta, \tag{5.7}$$

where the phase term $e^{-j\vartheta_k}$ represents the rephased signal. This further processing approach exploits the exact knowledge of the position of the target and assumes perfect synchronization among the network sensors. Even though it is impractical, however, it is very interesting from the scientific point of view because it provides an upper bound for the detection context as it maximizes the system SNR. Although the coherent approach is the most performing, it is also the most difficult to implement. As a matter of fact, to derive the exact coherent detectors defined in (5.6) and (5.7), a perfect synchronization between the network nodes has been assumed, with the raw information (i.e., amplitude and phase of signals) sent with no loss to the central processor unit. Unfortunately, there may be many cases where a perfect synchronization is impracticable or the raw data cannot be sent to the central unit due to the lack of a highly capable and reliable communication link. In these cases, a suboptimal processing strategy has to be implemented. This processing strategy is what we have previously indicated with decentralized processing.

5.3.2 Decentralized radar network processing

The decentralized network approach is considered suboptimum w.r.t. the centralized one. Decentralized processing approach is characterized by a double threshold for the overall detection stage. The use of this processing technique represents an alternative to the previous centralized approach for the overall radar networks processing strategies. This approach, even though can be considered a suboptimum processing, is also the most common practically adopted. It assumes that network nodes, in a first stage, work separately each other and all separated results are subsequently in the second stage fused together. In the first step, the "soft" detection (also indicated as soft decision) is performed by each individual sensor. In the second step, all the soft decisions are then jointly fused, thus providing a final "hard" decision output. The thresholds at all the nodes can be set to different values in order to guarantee f.i. the same false alarm rate (FAR) for all the sensors [22].

After each soft decision has been performed, a vector $\boldsymbol{v}$, containing only binary $\{0, 1\}$ values, zeros, if no detections are present, and ones if the target has been detected, are available for the subsequent processing stage. After the first stage, the hard detector decides for the presence of the target if at least Q elements of the vector v, are equal to 1, otherwise the detector will decide for the absence of the target. The number Q for the hard detection rule may vary according to the specifications of the system, such as for a specific value of the probability of false alarm or for a fixed value of the probability of detection, etc. The hard detector is indicated as follows [21]:

$$\sum_{k=1}^{NM} v(k) \geq Q. \tag{5.8}$$

Otherwise, when all the soft decisions coming from the overall distributed sensors have been collected, and accordingly to the detection criterion defined in formula (5.8), the hard detection rule is finally applied. Even if significantly different from the previous case, decentralized processing also provides the system with good jamming rejection capability [21]. As we previously mentioned, the centralized processing approach needs a perfect synchronization among the network sensors. On the other hand, the decentralized paradigm has less strict requirements and it can tolerate some synchronization inaccuracies. Moreover, another important difference between the two approaches is the link bandwidth required for transmitting the data from the nodes to the decision unit [23]. The centralized approach requires the transmission of both amplitude and phase of the overall received signals, the decentralized approach requires only the single subsystem decision instead (i.e., soft detection vector). The communication constraints for transmit and receiving the radar data could vary widely and, obviously, they should also be accounted for by the designer of the radar network system. Thanks to the abovementioned features, the decentralized network system is more robust respect to centralized system, and, therefore it is most commonly adopted in operative radar networks.

5.4 Radar network synchronization issues

As previously discussed, synchronization problems are among the main issues affecting almost every radar system component. They involve the transmitters, the receivers, and the target; they occur during the generation of the radar pulse, the internal communications of the radar system, and the electronic devices of the sensors themselves. As described in the previous section, the centralized radar network processing approach has the capability to maximize the SNR and it can easily be considered as the benchmark among all the detection processing approaches deployed in current operative radar networks. However, the centralized processing requires to ensure phase coherence for all the network elements [24]. In fact, phase synchronization is very important and represents one of the main differences between the centralized and decentralized radar processing approaches. Unfortunately, each element of a radar network usually operates with different local oscillators, and each of them is affected by statistically independent phase offsets. In other words, the transmitted carrier signal can oscillate by an unknown amount. Therefore, synchronization is mandatory for centralized processing. However, system synchronization determines additional hardware and software complexity (e.g., additional cost both in terms of computation time, further overhead for the coordination among the radar network elements, etc.). Anyway, given the potential benefits achieved by radar networks, the improved performance obtained by restoring synchronization is well worth the additional costs [24].

Network synchronization may be grouped into three macro areas: phase, frequency, and pulse synchronization issues [25].

From the point of view of the transmitter, in a centralized network system, phase and frequency synchronization are easily achievable since they can be easily measured and compensated for through the radar network nodes. Concerning pulse synchronization instead, in order to maximize as much as possible the received SNR, the successive received pulses should be added in such a way to provide the maximum amplitude signal. Therefore, pulse synchronization deals with the percentage of overlap between successive pulses as they are reflected from the target. If the different pulses are perfectly synchronized, the maximum SNR is achieved at the receiver side. Differently, if the pulses do not perfectly overlap each other, for example, due to changes in the synchronization status, the received SNR decreases.

Let us now consider the phase synchronization issue and how it affects the system SNR. Let us assume that N nodes simultaneously transmit one pulse. These pulses will reach the target at different time instants, depending on the relative distance between the sensor and the target and depending on the medium conditions in the observed area. For the sake of simplicity, let us consider just two received signals. As these pulses arrive at the receiver, they will combine according to the following model [25]:

$$\begin{aligned} r_1 &= A\sin\left(\frac{2\pi}{\lambda}\xi - vt\right), \\ r_2 &= A\sin\left(\frac{2\pi}{\lambda}\xi - vt + \varphi\right), \\ r_{12} &= r_1(\xi,t) + r_2(\xi,t) = 2A\cos\left(\frac{1}{2}\varphi\right)\sin\left(\frac{2\pi}{\lambda}\xi - vt + \frac{1}{2}\varphi\right), \end{aligned} \tag{5.9}$$

where ξ is the reference position variable, ν is frequency, t is the time variable, and φ is the phase difference between the two pulses. The phase difference between the transmitted pulses arriving at the target will lead to a reflected pulse with a different amplitude [25]:

$$|r_{12}| = 2A\cos\left(\frac{1}{2}\varphi\right) \tag{5.10}$$

and

$$\mathrm{SNR} = \frac{A'2}{kT} = \frac{\left(2A\cos\left(\frac{1}{2}\varphi\right)\right)^2}{kT}, \tag{5.11}$$

where A' is the reflected amplitude and kT is a scaling factor that takes into account the radar system disturbance. The SNR, described in formula (5.11), is characterized by a sinusoidal trend of the phase difference synchronization argument. Therefore, by varying the phase difference between the two pulses, it is possible to obtain an SNR gain equal to N^2 if the phase difference φ is an even integer multiples of π (perfect synchronization case). Otherwise, if the phase difference φ is an odd integer multiple of π the SNR will be 0 (completely destructive interference). Although a basic concept, it is vital to understand the importance of the topic of the synchronization issue.

As mentioned above, the point at which SNR must be maximized is upon reflection from the target to the receiver. The interaction between the transmitted pulses at the target is therefore the main concern. Several reasons exist why synchronization between pulses at the target might not occur in a manner that will provide a reflected pulse with the maximum possible SNR.

The first reason of possible mis-synchronization is related to the error made in the estimate of target's position. In this case, pulses will combine in such a way that the SNR is reduced, thus decreasing the overall radar network performance.

The second reason for synchronization loss is the timing error or, equivalently, the mis-synchronization between the times at which pulses are transmitted. The effect related to this error is that the transmitted pulses do not reach the target at the expected time, thus causing a loss in the SNR of the radar system.

The third cause of mis-synchronization is the Doppler effect, which is a subset of the timing error. In fact, pulses do not arrive at the target when they should since the Doppler effect, caused by the relative motion between the sensors and the target, brings to a shift in the measured position of a moving target. The result is that pulses lose some of the potential SNR gain, thus impairing the performance of the radar system network.

In order to deal with the aforementioned issues, the radar network must be able to employ the most suited algorithm accounting for the causes described above and compensating for the synchronization errors [25]. To obtain significant gains and to exploit the radar network capabilities, it is necessary to guarantee synchronization over time. A distributed radar network can reasonably be assumed to operate over a wide area with different weather conditions that might vary over time. The radar network elements have to interact with each other thus providing feedbacks and in order to rearrange themselves for modifying the operational conditions of the radar network. This adjustment has to be made with the intent of changing the system parameters in order to maintain phase and pulse synchronization. To accomplish the interaction between the network elements, it becomes necessary for these sensors to agree on a common notion of time. In centralized radar networks, the timing synchronization cannot be considered a big issue because the timing synchronization is typically managed by a common clock governed by the central processor unit and therefore there is no time ambiguity. In decentralized systems, there is no global clock or common memory. Each processor has its own internal clock and its own notion of time, and synchronization algorithms must be used [26].

The most important synchronization algorithms may be classified in two main families: the master–slave and the peer-to-peer. In the master–slave paradigm, one sensor is identified as the master node, and the other sensors as the slave nodes and each slave communicates only with the master node. The slave nodes consider the local clock reading of the master as reference time and they attempt to synchronize with the master. Instead, in the peer-to-peer paradigm any sensor can communicate directly with every other node in the network, thus eliminating the risk of a possible master node failure, which would prevent further synchronization issue. The peer-to-peer configurations offer more flexibility, but, at the same time, they are more difficult to control [27].

5.5 Data fusion methods

In order to exploit all the potentialities of radar networks, it is necessary to opportunely combine the information that the sensors collect over the area of interest. Data fusion is, thus, a very important issue for the realization of an operative radar sensor network. It is defined in literature as "the process of integration of multiple data and knowledge representing the same real world object into a consistent, accurate, and useful representation" [28]. The objective of data fusion is to combine relevant information from two or more sources into a single one in order to provide a more accurate description than any of the individual data sources [29]. In the mid-1980s, the Data Fusion Group proposed a conceptual model that identified the processes, functions, and categories of the specific techniques applicable in the framework of sensor data fusion. Currently, there are six levels composing the data fusion model. They are [30]:

- Level 0: source preprocessing/subject assessment. An initial process allocates data to appropriate processes and performs data pre-screening.
- Level 1: object refinement. This process combines position, parametric, and identity information to achieve refined representations of individual observed targets (e.g., emitters, platforms, weapons). Level 1 processing performs four crucial steps: (1) transforms data from multiple sensors into a consistent set of units, (2) refines the estimation of the position and kinematics parameters of

targets, (3) assigns data to objects thus allowing the statistical estimation techniques, and (4) performs the classification of the observed target [28].

- Level 2: situation refinement. Level 2 processing develops a description of current relationships among objects and events in their environmental context. Situation refinement addresses also the data interpretation in an analogous way how a human might interpret the meaning of the sensor data [28].
- Level 3: impact assessment: critical refinement (or threat refinement). Level 3 processing projects the current situation into the future to draw inferences about threats and opportunities for operations [28–31]. Among the possible strategies, expert systems, blackboard architecture, and fuzzy logic techniques can be applied.
- Level 4: process refinement. Level 4 processing is a sort of meta-process, it controls other inner processes. Level 4 processing, as for level 1, performs four key functions: (1) it monitors the performance of the data fusion process in real time, (2) it identifies what information is needed to improve the multilevel fusion product, (3) it determines the source specific requirements to collect relevant information (i.e., which sensor type, which specific sensor, which database, etc.), and (4) it controls, allocates, and directs the sources to achieve the overall system objective. Level 4 processing is also considered as a borderline data fusion process as it is partially inside and partially outside the data fusion context [28–32].
- Level 5: user refinement (or cognitive refinement). Level 5 processing performs a data adaptation between who queries information and who has access to the information in order to support cognitive decision-making and actions (i.e., human–computer interface) [33].

5.5.1 Data fusion architectures in multistatic radar networks

One of the main concerns in developing an operative radar network is where, in the data flow, to actually fuse together the data streams coming from all the sensors. The architecture of a radar network is closely related to the type of data fusion that is performed. Typically, there are three alternatives for the fusion system structure [34]:

- Centralized
- Decentralized
- Hierarchical or hybrid.

In a centralized fusion paradigm, data fusion is handled by a central processor unit (or data fusion center, see Figure 5.6), to which each element of the network is interconnected. The fusion center collects all the information from all radar nodes and decides the actions performed by each sensor.

In the decentralized data fusion paradigm, data are fused directly at the local sensors by local processing units (see Figure 5.7). In this case, every radar sensor can be considered as an intelligent system having some degree of autonomy in decision process. Sensor coordination is achieved by connecting network nodes

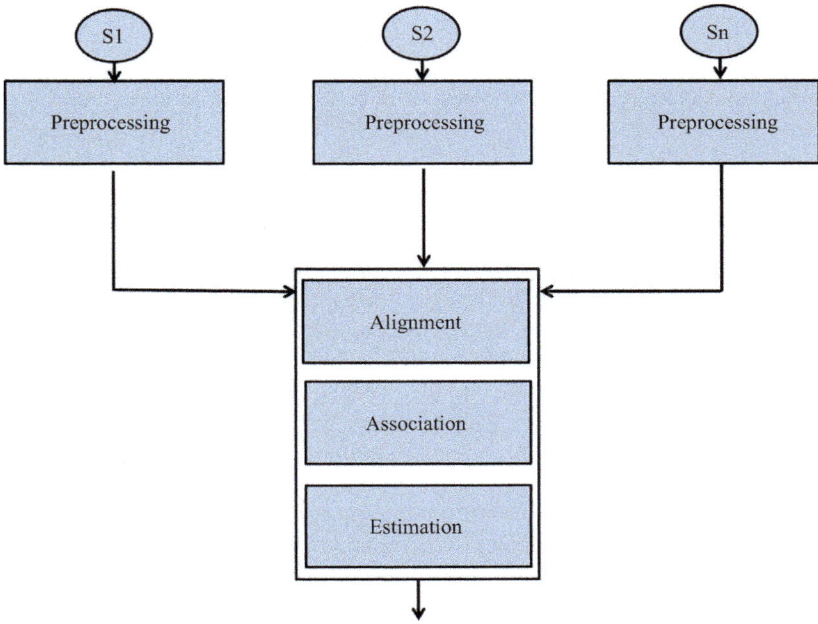

Figure 5.6 Centralized fusion paradigm

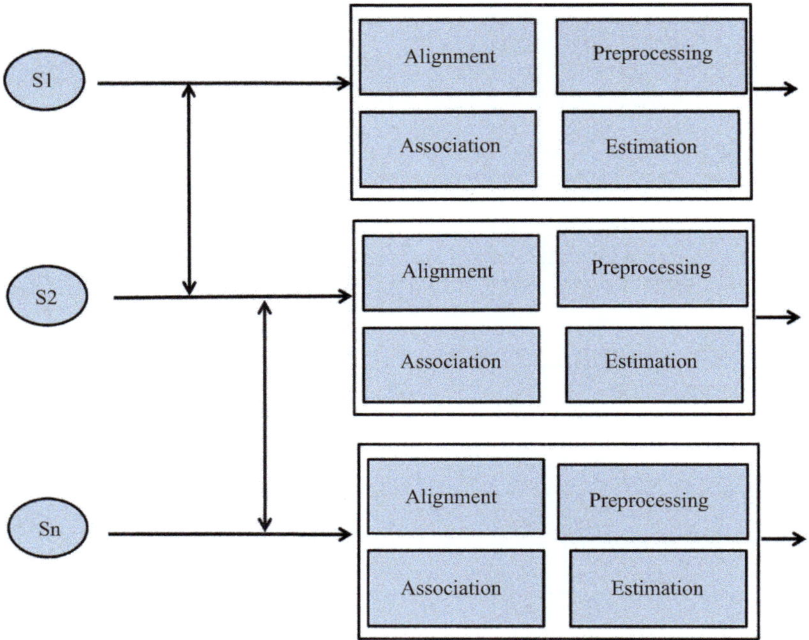

Figure 5.7 Decentralized fusion paradigm

through communication links. In [35] it has been observed that with the decentralized data fusion paradigm it is possible to achieve the following benefits:

- structure scalability, with no constraints in terms of computation and communication limitations;
- survivability in the case of communication link failures, or for possible dynamic changes of the network system; and
- modular approach for designing and implementing the fusion modules.

However, the amount of information shared among the sensors could be a serious problem that may arise in decentralized fusion networks [36]. Moreover, another very important aspect is the data exchange management among the nodes in the absence of a common control facility. In fact, a delay between senders and receivers could infer a transient inconsistency of the network status, thus leading to performance degradation of overall network system [37].

Finally, the hierarchical/hybrid fusion approach consists of both centralized and decentralized fusion structures. In a hybrid fusion system, there are usually different levels of hierarchy in which the highest level is represented by a global fusion center, while the lowest level consists of multiple local fusion centers that are all interconnected with the global fusion center [38] (see Figure 5.8). Each local center is responsible for the management of a subset of network nodes.

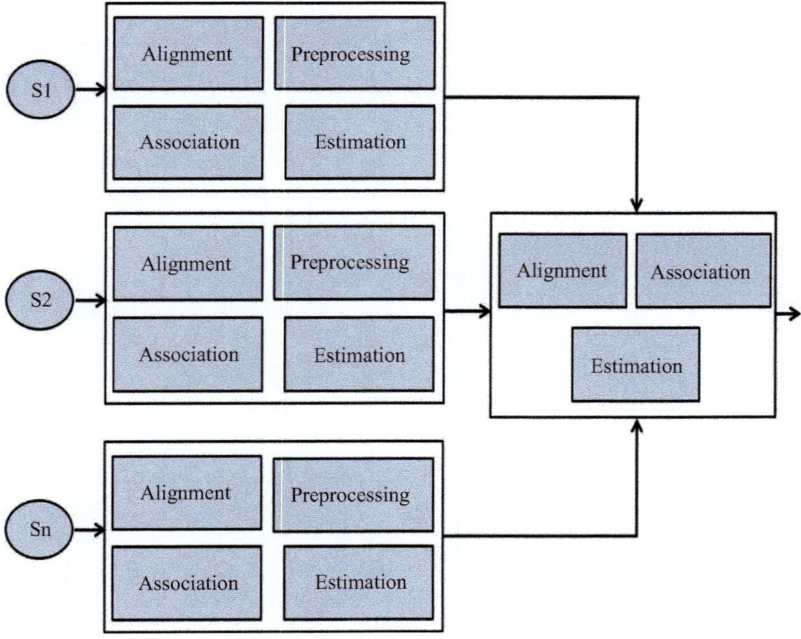

Figure 5.8 Hybrid fusion paradigm

The partitioning rule can be realized on the basis of geographical locations paradigm or the sensor function performance.

5.5.2 Information fusion approaches in multistatic architectures

There are three main approaches for fusing radar network information [28]. The first approach is the fusion of the raw data observed by each radar node, while the second approach is the fusion of parametric state vectors, which typically represent an optimum estimate achieved by each individual sensor measurements (e.g., the target detection, the target position, and the velocity of the target). The third approach is the hybrid fusion approach, which exploits either raw data or processed state vector data.

The fusion of raw data is typically performed in the centralized architecture paradigm (centralized fusion approach). First, in a centralized architecture framework, the raw data coming from the different sensors network are typically aligned to common coordinates and units in order to be easily processed by the central processing unit. Data are then associated to determine the correspondence between the sensor and the observations. That is, in a radar network environment, it is necessary to determine which observations are representative of the same target. This association issue may be very complicated, especially in environments dense of targets. Nevertheless, once the association phase is terminated, then the data are typically fused by using sequential estimation techniques, such as Kalman filter-based algorithms [39]. This centralized fusion approach is theoretically the most accurate way to fuse data. However, this approach requires also the transmission of raw data from all the network nodes to the central processing facility. Hence, in the case of large amount of data (e.g., imaging data), this may require very large communication bandwidth, which would exceed what is available with the current technology in a real-time operational context.

The second information fusion approach is commonly implemented in a decentralized architecture paradigm (i.e., distributed fusion approach). In fact, for decentralized architectures each sensor typically performs single-source positional estimations, thus producing information about target state vectors for each network node. This means that each sensor provides an estimate of the position and velocity of the detected targets, based only on its own single-source data. These estimates of position and velocity (i.e., a state vector estimate) are then considered as the input for the data fusion process in order to achieve a joint state vector estimate that is based on the multiple sensors output. It should also be remarked that in this approach the data alignment function and the data association procedure still need to be performed, but are now performed at the state vector level rather than at the raw data level. The main important advantages of the decentralized fusion approach w.r.t. the centralized approach are essentially related to the reduced volume of data that is exchanged among the sensors. In fact, the sensing data, collected by each network sensor, are compressed into a representative state vector that typically has a limited computational workload w.r.t. the raw data sensing. In addition, for the decentralized fusion approach the association process is conceptually easier than that performed at the raw data level [28]. However, the decentralized fusion approach is not as accurate as centralized fusion because there is an information loss between the sensors and the fusion process. In particular, raw data contain more details and information about targets.

A last comment is dedicated to hybrid information fusion that combines raw data and state vector fusion approaches. This approach tries to combine the advantages, while limiting the shortcomings, of both approaches. Indeed, during ordinary operations, the state vector fusion approach is typically performed in order to reduce the data volume and communications demands. Conversely, when more accuracy is required, or the radar network is in a target-crowded scenario, the system may decide to adopt a centralized fusion approach. Alternatively, in some extreme cases, a combination of state vectors and data may be fused. Even though the hybrid architecture may provide very good flexibility, it also requires additional processing overhead to monitor the fusion process and consequently select between the centralized and decentralized fusion approaches.

There is no single optimal architecture for any given information fusion applications. Instead, the choice of architecture and data fusion strategy must balance computing resources, available communication bandwidth, desired accuracy, and sensors' capabilities [40].

5.5.3 Achievements in radar network fusion

Radar network technology and multistatic data fusion are two aspects of a same problem that is in continuous evolution. As a matter of fact, currently there is a significant amount of ongoing research that is devoted to develop new techniques, to improve existing algorithms, and to understand how to assemble these aspects into an overall network architecture able to provide multiple methodologies of data fusion [28]. By considering the Joint Directors of Laboratories/Data Fusion Information Group data fusion model introduced in the previous section, the most developed area of data fusion is Level 1 processing.

In the radar network fusion context, the problem of detection, tracking, and target imaging, based on multiple sensor observations, is a relatively old problem. As matter of fact, Gauss and Legendre, in the eighteenth century, developed the first least-square method for estimating the orbits of asteroids [41]. Many numerical techniques exist to detect targets and to estimate their position and velocity. The Kalman filter [42] is the most famous and utilized sequential estimation technique. Currently, the issue of single target tracking in high SNR environments and for dynamically predictable targets can be considered straightforward. Most of the scientific research attention is now focusing to solve the maneuvering target problem in a multitarget and multisensor context, in the presence of complex scenarios involving, for example, multipath propagation effects, co-channel interference, or clutter disturbance. As a fact, in the last decades, numerous innovative techniques such as multiple hypothesis tracking (MHT), probabilistic data association (PDA) methods, random set [43], and multiple criteria optimization theories [44] were developed to solve the multisensor MTT issue. Recently, the aforementioned techniques have been simultaneously used by selecting the most appropriate solution based on algorithm performance and operative context [28].

Level 2 and 3 approaches are still relatively immature respect w.r.t. level 1, with very few robust operational systems. Concerning level 2 processing, Bayesian decision theory and the Dempster–Shafer Evidential Reasoning [45] are the most common techniques employed. On one hand, Bayesian decision theory is used to generate a probabilistic model of uncertain system states by consolidating and interpreting overlapping data provided by several radar sensors [28]. On the other hand, the Dempster–Shafer method is considered an alternative to traditional probabilistic decision theory [46]. In fact, in such a method hypotheses do not have to be mutually exclusive, and the probabilities involved can be either empirical or subjective [47]. Currently, the main methods for level 3 fusion processing are expert systems [48], blackboard architecture [49], and fuzzy logic [50].

Level 4 assesses and optimizes the performance of an operative data fusion process. However, modern radar networks involve multiple sensors, operative mission constraints, dynamic observing environments, multiple targets, etc. Therefore, level 4 processing level is not yet scientifically developed and it represents an open research topic.

5.6 Multitarget tracking

The problem of MTT has a tradition of over 40 years of studies, see [42]. The MTT problem was well explained by Jeffrey Uhlmann [51], who described it using an example derived from baseball game. When a Major League outfielder runs down a long fly ball, the tracking of a moving object looks easy. Over a distance of a few hundred feet, the fielder calculates the ball's trajectory to within an inch or two, and times its fall to within milliseconds. But what if an outfielder were asked to track 100 fly balls at once? As it turns out, even 100 fielders trying to track 100 balls simultaneously would likely find the task an impossible challenge. Problems of this kind do not arise in baseball, but they have considerable practical importance in other realms, such as ballistic missile defense, visual surveillance [52], biomedical analytics [53,54], robotics [55,56], assisted living [57,58], and autonomous driving [59–61].

A fundamental aspect of MTT, already recognized in 1964 [62] by Sittler, is the measurement origin uncertainty (MOU) [63–67], that is, a measurement, used in the tracking algorithm to update the target track, might not have originated from the real target of interest. This situation can occur in surveillance systems, where sensors are typically operating in a cluttered environment, in which the FAR is high. This problem can also happen when several targets are in the same neighborhood and it is not possible to associate with certainty the observed detections (assumed that they have been resolved) with the various targets to yield the final measurements. A similar situation can occur in the track formation problem when there are several targets but their number is not known and some of the measurements might be spurious. In [68] the authors computed a scalar information-reduction factor due to the MOU for the single target case quantifying the degradation of estimation performance in the presence of MOU. Indeed, one of the main differences between filtering and MTT is the MOU. In fact, in simple filtering techniques, the measurements are not ambiguous, that is, they are perfectly associated with the target. Then, the application of standard estimation algorithms, which use, for instance, the nearest neighbor filter, can lead to very poor results in environments where spurious measurements frequently occur [69]. One of the first successful approaches, proposed by Bar-Shalom in [63,65,66], to cope with the MOU problem was the PDA for the single target case and its generalization to the multiple target case, the Joint Probabilistic Data Association (JPDA). An overview of the PDA technique and its application to different target tracking scenarios is given in [70]. At the same time an alternative and powerful algorithm, the MHT, was proposed. The basic idea of propagating multiple hypotheses for a single target in a false alarm background was given in [64] while Reid [67] first developed a complete algorithmic approach. In [71] Blackman provided an overview of the basic principles behind the MHT and the alternative implementations for use in common cases. Both the JPDA and the MHT have been fundamental building procedures for the future development of MTT strategies, leading to more and more sophisticated solutions, as detailed in [70,71]. Among the most important developments we include the interacting multiple model (IMM) filtering, see, for example, [72,73], to track targets during maneuvers, the variable structure IMM [73] to incorporate prior information in the tracker, the capability to deal with unresolved targets, see, for example, [74], the random matrix framework in order to track extended targets. A quantitative comparison among the MTT strategies is provided in [75] and [76], where the techniques have been categorized into more than 35 different algorithmic typologies. The comparison is provided in such a way that lists each algorithm and categorizes the processing scheme, data association mechanism, complexity scaling (with number of targets and with state dimension), overall complexity, and a subjective performance figure. MTT methods, like the JPDA and MHT, are based on conventional probability theory [77]. A more recent class of MTT methods is based on finite set statistic [78–81]. Here, target states and measurements are modeled as random finite sets (RFSs), which means that they have no order and also the number of elements in the set is modeled as an unknown random variable. The main advantage and attractive characteristic of the RFSs are provided by Mahler seminal work [78]. A Bayesian filtering solution, the probability hypothesis density (PHD) filter, is proposed in [75], which the first-order statistical moment of the multitarget posterior is propagated instead of the full posterior distribution. The PHD is the function whose integral in any region of state space is the expected number of targets in that region. The PHD filter and its extension, the cardinalized PHD (CPHD) [82], are gaining increased popularity in the MTT community and led to many different derivations, interpretations, and implementations [79]. In [81] the asymptotic characterization of the PHD in the limiting regime where the number of sensors goes to infinity is studied, proving that the PHD is asymptotically efficient. Further developed in the RFSs framework include the multi-Bernoulli filter [83] and the labeled multi-Bernoulli filter [84,85]. Common building blocks useful in both the classical MTT and RFSs based MTT are the sequential Monte Carlo (SMC) methods [86,87]. The SMC methods provide computational efficient solution for sampling probability distribution of time-varying random objects. They enable the development of MTT algorithms using arbitrary nonlinear non-Gaussian motion and measurement models. Many MTT methods are computationally demanding and their complexity does not scale well with the number of targets and other relevant system parameters. Thus, they are often impractical for use on resource-limited devices. MTT with low complexity and good scalability can be obtained by using the methodology of belief propagation (BP). BP provides a principled approximation of optimum Bayesian inference that achieves a very attractive performance-complexity compromise [88,89]. Due to its generality and flexibility, it is suited to general nonlinear, non-Gaussian system models and it is able to accommodate unknown and time-varying hyperparameters. However, only recent works have considered its use for MTT [90–96]. BP-MTT is a promising methodology because it yields a highly efficient solution for the data association problem. Because of their low complexity and good scalability, BP-based methods are also suitable for large-scale tracking scenarios involving a large number of targets and/or sensors and/or measurements, and for use on resource-limited devices. In the recent years, such MTT methods have been successfully applied to real-world surveillance scenarios using radar technologies, as described in [97,98].

5.6.1 Description of the MTT problem for radar network

The MTT problem, in its most general form, can be described as follows. At (discrete) time k, there are N_k targets, denoted by the index $i \in \{1, 2, \ldots, N_k\}$. The state $x_k^{(i)}$ of the ith target at time k, described by a random vector, consists of the position of the target and possibly further parameters (i.e., velocity). In the case of multiple radars $s \in \{1, \ldots, S\}$, at time k the radar network generates $M_{k,s}$ measurements $z_{k,s}^{(m)}$, with $m \in \{1, \ldots, M_{k,s}\}$.

Furthermore, for each sensor, a measurement cannot originate from more than one target simultaneously. Therefore, one target can generate at most one measurement per sensor. The probability that a measurement originates from target i (i.e., sensor s detects target i at time k) is $p_d^{(s)}$. Such detection probability can be dependent on the electromagnetic environment, see, for example, [15,99,100]. Each sensor collects also measurements originated from clutter, which are often modeled as a Poisson random variable with parameter $\mu_c^{(s)}$. At time k the single-sensor likelihood function can be then obtained, using the standard MOU assumptions [42] and [95,96]:

$$
\begin{aligned}
f\left(z_{k,s}, M_{k,s} | \boldsymbol{x}_k\right) = & \frac{e^{\mu_c^{(s)}}}{M_{k,s}!} \left(1 - p_d^{(s)}\right)^{N_k} \prod_{M=1}^{M_{k,s}} \mu_c^{(s)} f_c^{(s)} z_{k,s}^{(m)} \\
& \times \sum_{a_{k,s}} \psi\left(a_{k,s}\right) \prod_{i \in N_{a_{k,s}}} \frac{p_d^{(s)} f^{(s)}\left(z_{k,s}^{(m)^i} \middle| x_k^{(i)}\right)}{\mu_c^{(s)} f_c^{(s)}\left(z_{k,s}^{(m)^i}\right)\left(1 - p_d^{(s)}\right)},
\end{aligned}
\tag{5.12}
$$

where $x_k \triangleq [x_k^{(1)T}, \ldots, x_k^{(N_k)T}]^T$, $f^{(s)}(\cdot)$ and $f_c^{(s)}(\cdot)$ are respectively the target-originated likelihood and the clutter distribution, $a_{k,s}$ is the association vector between measurements and targets (see details in [95,96]), $\mathfrak{R}_{a_{k,s}}$is the set of observed targets by sensor s at time k, $\psi(a_{k,s})$ is an indicator function that is zero when the associations $a_{k,s}$ for a given target state $\boldsymbol{x}_k$ are unfeasible. Stated that the number of associations grows exponentially with both the number of targets N_k and the number of measurements, computation with brute force is, thus, often unfeasible without proper "approximations." The MTT problem from a statistical point of view is twofold in the sense that it contains both a detection problem and an estimation problem at the same time [81]. The MTT problem can be summarized as follows. Based on the observations $z_{k,s}$, $s = 1, \ldots, S$ available up to time k, we want to:

1. Detect the number of targets $\hat{N}_k$.
2. Estimate their state $\hat{x}_k \triangleq [x_k^{(1)}, \ldots, x_k^{(N_k)}]^T$.

Assuming to know (or estimate) the posterior distribution of $\boldsymbol{x}$, then optimal solutions would involve an optimization procedure based on the minimization of a Bayesian cost. However, even when the posterior is available such procedure risks to be computationally demanding. Most of the MTT procedures try to approximate the posterior and its related estimators. For instance, the JPDA, provided the number of targets by the track logic, assumes that the marginal posterior (and prior) of each target is Gaussian and then computes its mean and covariance [42]. The PHD is the function representing the expected value of the RFS containing the element of $\boldsymbol{x}_k$ [78]. The BP tracker [95] and [96] computes an approximate marginal posterior of each element of $\boldsymbol{x}_k$. Performance metrics of the MTT problem can be related only to the detection problem, for instance, see the Time-on-Target (ToT) versus FAR [100], or only to the estimation problem, for instance, see the mean square error (MSE) [15]. The optimal subpattern assignment, proposed in [101], is a metric that takes into account both the detection and the estimation errors.

5.7 Radar networks for maritime surveillance: a recent experimentation

In this section, we describe a recent experimentation conducted by the Centre for Maritime Research and Experimentation of the NATO Science and Technology Organization. In particular, a simple network composed by two oceanographic HFSW radars, precisely two Wellen RAdar (WERA) systems developed at the University of Hamburg [102], was deployed along the Italian coast of the Ligurian Sea (Mediterranean Sea) [97,103] for remote sensing and for maritime surveillance.

5.7.1 Experimental setup

During the Battlespace Preparation 2009 (BP09) campaign, two WERA systems were deployed on Palmaria island close to the city of La Spezia (44°2′3″ N, 9°50′36″ E) and at San Rossore park near Pisa (43°40′53″ N, 10°16′52″ E). A simple picture representing the experimentation, together with the HFSW radar coverage, is given in Figure 5.9.

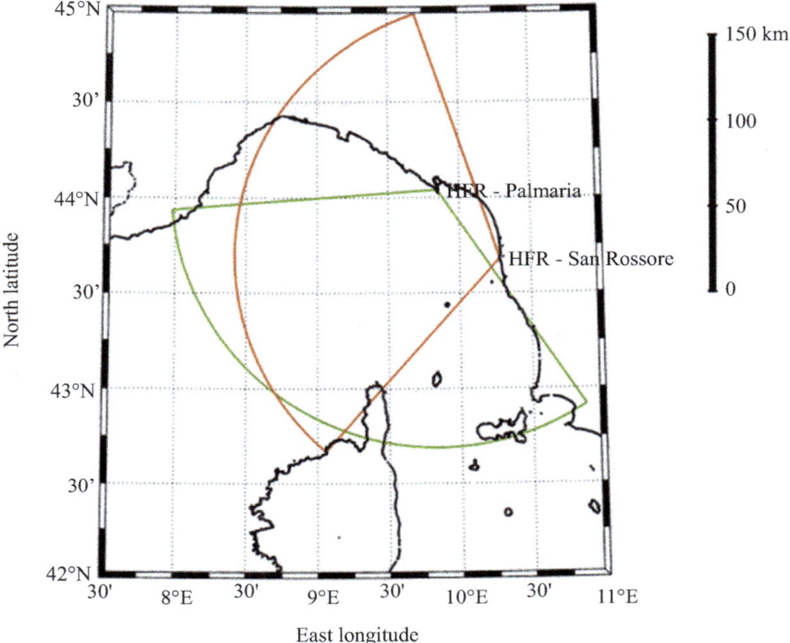

Figure 5.9 NATO experiment on HF radars: (green) Palmaria and (red) San Rossore sites and covered areas [103]

Compared with the very large distances illuminated by the two sensors (well beyond the optical horizon), WERA can be considered to be a quasi-monostatic system, even if the transmitter (*Tx*) and receiver (*Rx*), made up with $\lambda_0/4$ dipole arrays (where λ_0 is the transmitted wavelength), are separated by approximately 300 m. In the ground-based installation setup, the transmitter has typically a rectangular arrangement, while during the experiment, the receiver was made by a 16-element linear array (different *Rx* array configurations are available), with a separation of $\lambda_0/2$ between the array elements. The angles w.r.t. north of the two array installations were 296.2° and 12.0°, respectively. The two WERA systems transmitted a linear frequency-modulated-continuous-wave signal using the same operating frequency (i.e., $f_0 = 12.5$ MHz), but with orthogonal waveforms (i.e., upsweep and downsweep). The range resolution was 1.5 km, with chirp bandwidth $B = 100$ kHz. The average power transmitted by both radar systems was approximately 35 W. The processing chain was as follows. Target detection was performed by means of the 3-D (i.e., range, azimuth, and Doppler) ordered statistics constant FAR (OS-CFAR) algorithm, developed by the research group of the University of Hamburg [104]. Coherent processing intervals, not statistically independent, were made of 512 (or 256) samples with an overlap of 75%, that is, with detections occurring every 33.28 s (or 16.64 s). The two systems were made synchronous by means of a GPS clock. It is important to observe that such long integration intervals are required for oceanographic parameter estimation in the HF band. In fact, the two systems were initially deployed mainly for remote sensing purposes (i.e., sea state, wave spectrum, wind and sea current sensing). Ship detection and tracking capabilities were made possible at the cost of an increased computational complexity. Automatic identification systems (AIS) ship reports were provided by the base station located at Castellana (44°4′3″ N, 9°48′58″ E, at 200 m altitude). As it will be made clearer in the following, these data were used as ground-truth information for assessing system performance.

5.7.2 Performance assessment

In this section, we describe the procedure adopted for assessing the performance of the target detection, tracking, and fusion algorithms. Usually, ships and vessels exceeding a given gross tonnage[1] are equipped with AIS transponders for tracking and reporting their position to nearby vessels and to ground-based stations. As mentioned, the AIS information was used as ground-truth data. The MTT strategy was based on the JPDA paradigm, that is, a Bayesian approach that associates all the validated measurements to the tracks by using probabilistic weights. The track initiation/confirmation and track termination logics were respectively based on the *M* out of *N* (*M/N*) and *N**/*N** criteria. The unscented Kalman filter (UKF) was applied, thus providing a good trade-off between tracking performance and algorithm computational burden. Indeed, the use of range-rate measurements, which are correlated to the range measurements, would recommend to work with the UKF, instead of using the extended Kalman filter or converted measurement-Kalman filter. The confirmed tracks generated by the MTT at each HFSW radar site were then combined by means of a track-to-track association and fusion (T2T-A/F) logic described in [97]. The detection and tracking capabilities of the proposed MTT strategy and the final T2T-A/F system were quantitatively evaluated by adopting the following metrics:

- ToT and FAR: The ToT is defined as the ratio between the time during which the tracking algorithm follows the target and the whole interpolated ship route (recalling that AIS data occur at times not synchronized at all with the radar timestamps). The FAR is defined as the number of false detections and track contacts, normalized by the recording time interval and the area of the all surveyed region. An ideal system would have a $ToT = 100\%$ with $FAR = 0$ (i.e., no false alarms).
- RMSE: The accuracy of the radar system is evaluated in terms of the RMSE and represents the localization capacity of a detected target.

[1]The AIS is required for all the ships exceeding 300 gross tonnage and engaged on international voyages for all cargo ships of 500 gross tonnage and not engaged on international voyages and for all passenger ships.

5.7.3 Experimental analysis

The OS-CFAR, JPDA-UKF, and T2T-A/F procedures were tested on a 25-day data set collected between 8 May and 4 June 2009. The analysis was performed only in the fusion region (i.e., the region in which sensors' coverage areas overlap), as reported in [103]. In this way, it was possible to observe the same ship routes and to provide a fair comparison between the single radars and the fusion system. In this region, the average number of AIS-carrying vessels per day varied between 59 (May 18) and 91 (May 26). The parameters used in the MTT algorithms are summarized in Table 5.1, while the parameters used for performance assessment are given in Table 5.2.

A sample 1-day output is depicted in Figure 5.10. The ship trajectories and the true active tracks are displayed as follows. The output tracks of the T2T-A/F and the AIS ship routes are depicted in the surveyed area with the blue and gray lines, respectively. The two JPDA-UKFs, Palmaria (green) and San Rossore (red), are reported as well. As we can observe from a preliminary visual comparison, there is a quite good agreement between the output tracks and the AIS available ship routes up to about 100 km distance from the coast.

Table 5.1 MTT parameters

Specification	**Value**	**Parameter**
Sampling period	16.64 s/33.28 s	T_k
Process noise	$1 \cdot 10^{-2}$ ms^{-2}	σ_v
Standard deviation range	150 m	σ_r
Standard deviation bearing	1.5°	σ_b
Standard deviation range-rate	0.1 ms^{-1}	σ_r
Detection probability	0.35	P_D
Clutter density	10^{-9} ms^{-2}	λ
Gate threshold	3.3^2	γ
Init. filter (pos.)	500 m	$\sigma_{x,y}$
Init. filter (vel.)	10 ms^{-1}	$\sigma_{x,y}$
Maximum velocity	25 ms^{-1}	v_{max}
Track initiation logic	7/8	M/N
Track termination logic	3	N^*

Table 5.2 Performance assessment procedure parameters

Specification	**Value**	**Parameter**
PVR range threshold	1.5 km	δ_r
PVR bearing threshold	2°	δ_b
PVR range-rate threshold	2 ms^{-1}	δ_r
Maximum time for AIS flag	30 s	δT_{max}

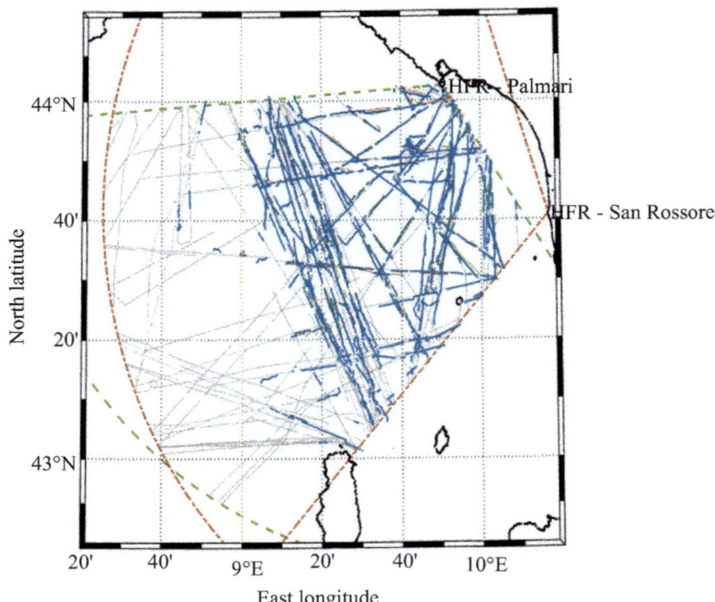

Figure 5.10 True active tracks in the fusion region: (black) AIS data, (green) tracking routes relative to the Palmaria sensor, (red) San Rossore system, and (blue) fusion T2T algorithm for the network radar system [103]

5.7.3.1 ToT analysis

The ToT analysis is presented in Figure 5.11 at the varying of range, azimuth, and range-rate, respectively. The ToT versus range is shown in Figure 5.11(a) and (d), is estimated over 10 km distance intervals, and averaged over the whole azimuth and range-rate intervals. The peak values w.r.t. Palmaria occur in the 10–50 km interval, where the ToT is about 65%–77% for the JPDA-UKF (red) and 49%–58% for the OS-CFAR algorithm (yellow), while it is about 80%–90% for the T2T-A/F strategy (blue) (see further details in [103]). The maximum improvement of the JPDA-UKF w.r.t. the OS-CFAR is 20%, while it is 15% for the T2T-A/F w.r.t. the JPDA-UKF. Beyond the 80 km limit, the T2T-A/F relies almost completely on the JPDA-UKF output of Palmaria. Performance rapidly decreases, and the differences among the curves become negligible.

For San Rossore, the peak ToT value is about 55%–77% for both the tracker and the detector. However, on average, the OS-CFAR achieves better performance than the JPDA-UKF. Since the two sensors share the same setup parameters and observe the same vessels, a possible reason can be found in the relative geometry between the ship routes and the sensor positions, supposedly more favorable for Palmaria than for San Rossore. It is important to observe that no AIS reports were available in the first 10 km range from San Rossore. Finally, the data fusion

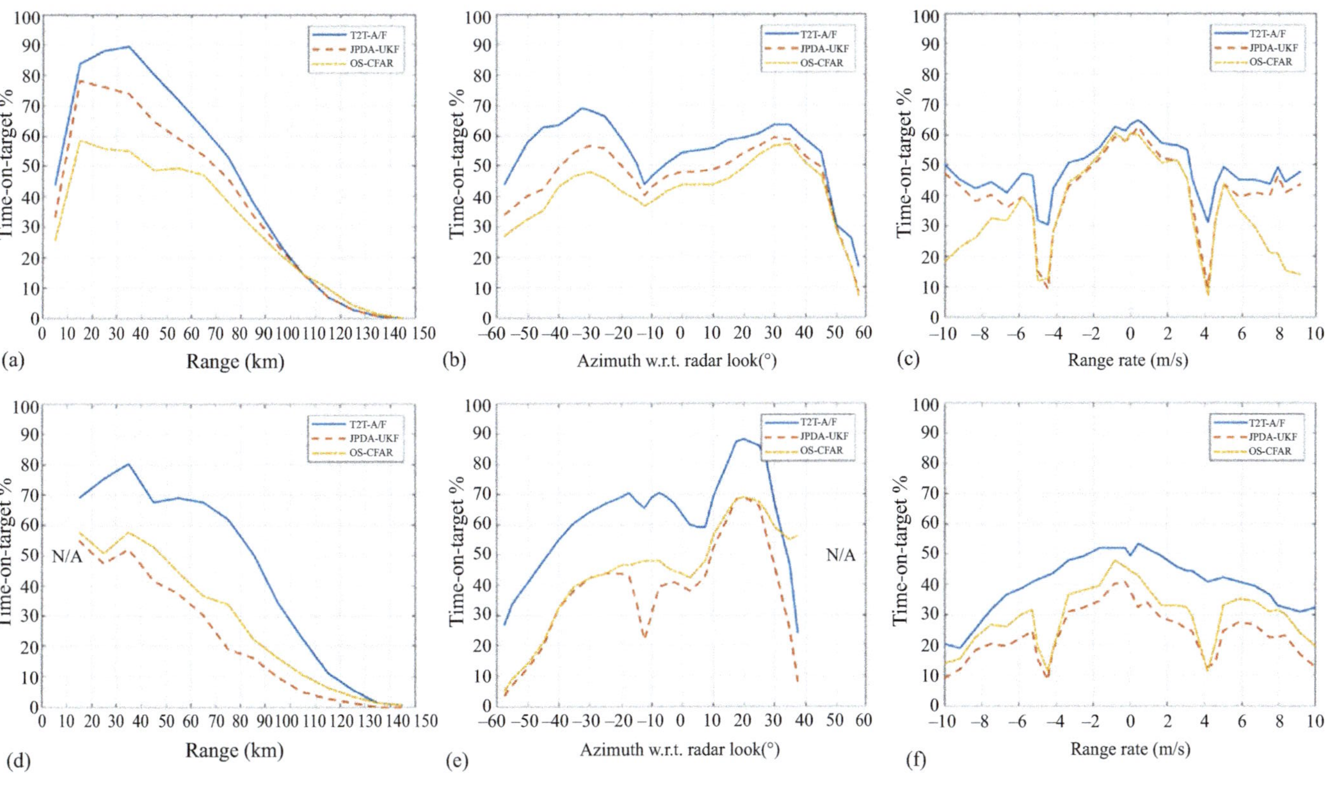

Figure 5.11 Estimated ToT percent versus range (km), azimuth (°), and range rate (m/s), w.r.t. (upper plots) Palmaria and (lower plots) San Rossore sites: (blue) T2T-A/F, (red) JPDA-UKF, and (yellow) OS-CFAR. (a) ToT versus range as regards Palmaria. (b) ToT versus azimuth w.r.t. Palmaria. (c) ToT versus range rate as regards Palmaria. (d) ToT versus range w.r.t. San Rossore. (d) ToT versus azimuth as regards San Rossore. (f) ToT versus range rate w.r.t. San Rossore [103]

strategy leads to a final ToT value of 79%, with a maximum improvement of about 30%w.r.t. to the JPDA-UKF.

5.7.3.2 FAR analysis

As for the ToT, the FAR analysis is carried out at the varying of range, azimuth, and range-rate and averaged over all the days, on the same intervals considered for the ToT (see further details in [103]). On each specific interval, FAR values are normalized such that their weighted sum provides the total FAR per unit of time and area for that day. The FAR versus range is shown in Figure 5.12(a) and (d). As expected, the application of the tracking algorithm significantly reduces the number of false track contacts and cancels most of clutter originated returns (cfr. red (JPDA-UKF) and yellow (OS-CFAR) lines), particularly at far distance. False detections manifest a more uniform behavior along range than the tracker outputs, which tend instead to accumulate in the first 80 km from the radars.

The distribution of the false contacts versus azimuth is depicted in Figure 5.12(b) and €. At Palmaria, no significant differences arose from the analysis of the JPDA-UKF and T2T-A/F outputs, except for an increase at angles smaller than $-15°$. This is in agreement to what is shown also in Figure 5.12(b). At San Rossore instead, the increase in the FAR is almost constant along azimuth. As expected, an increase in the ToT estimate corresponds to an increase in the FAR, as shown in Figure 5.12(e). The estimated FARs versus range rate are shown in Figure 5.12(c) and (f) for Palmaria and San Rossore, respectively. A relevant amount of false contacts are produced by sea clutter, not perfectly filtered out by the OS-CFAR detection algorithm (yellow line).

5.7.3.3 RMSE analysis

The RMSE is computed as regards track length for both the position and the velocity components of the target state vector. Results are shown in Figure 5.13.

The blue curves were obtained averaging all the sub-tracks up to 100 samples (i.e., corresponding to about 1 h length) of the T2T-A/F system. The green and red curves are obtained from the parent tracks of Palmaria and San Rossore, respectively. Let us consider the RMSE of the position estimate (see Figure 5.13(a)). The error spans between 0.6 and 1.0 km, for both Palmaria and San Rossore. As expected, the errors of the two stand-alone systems are close, while the T2T-A/F system provides an RMSE significantly below that of the two JPDA-UKFs, about 200–300 m on average. The RMSE of the velocity estimate is presented in Figure 5.13(b). Between the T2T-A/F output and the two trackers, significant differences arose, not only in terms of mean error level (about 0.5 m/s smaller) but also in terms of transitory error, almost eliminated by the T2T fusion algorithm. In fact, when one of the two sensors loses its track, the other one most likely is able to follow it.

5.7.3.4 Final remarks

In this brief excursus on a real-operative radar network, the overall system performance was compared to single-sensor performance. First of all, a comparison between the JPDA-UKF tracking algorithm and the OS-CFAR detector was

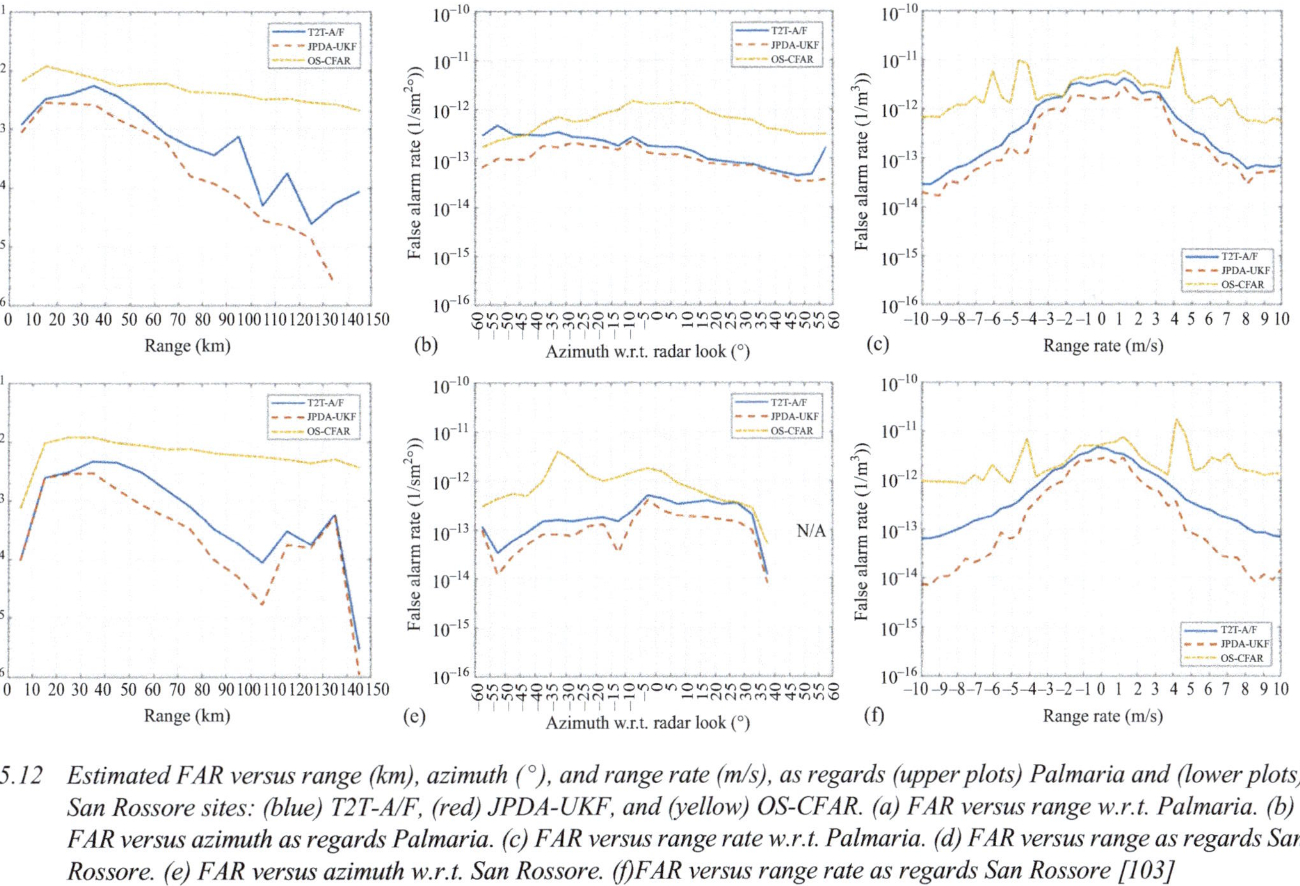

Figure 5.12 Estimated FAR versus range (km), azimuth (°), and range rate (m/s), as regards (upper plots) Palmaria and (lower plots) San Rossore sites: (blue) T2T-A/F, (red) JPDA-UKF, and (yellow) OS-CFAR. (a) FAR versus range w.r.t. Palmaria. (b) FAR versus azimuth as regards Palmaria. (c) FAR versus range rate w.r.t. Palmaria. (d) FAR versus range as regards San Rossore. (e) FAR versus azimuth w.r.t. San Rossore. (f)FAR versus range rate as regards San Rossore [103]

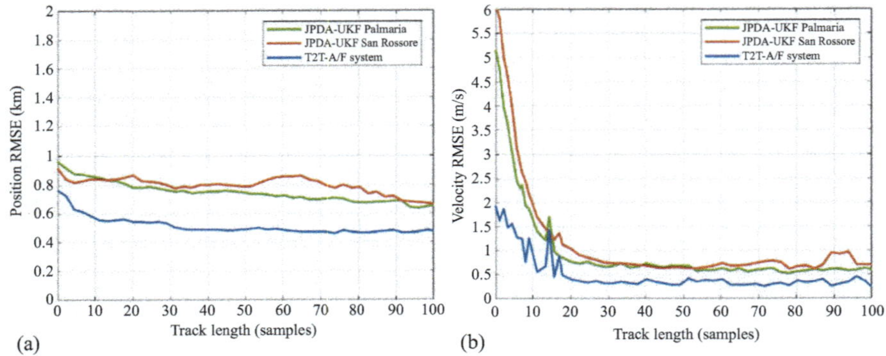

Figure 5.13 RMSE of the (a) position and (b) velocity state vector components: (green) Palmaria JPDA-UKF, (red) San Rossore JPDA-UKF, and (blue) T2TA/ F; (a) RMSE for position and (b) RMSE for velocity [103]

carried out. In terms of average ToT, as expected, both the detector and the tracker achieved similar results. The T2T-A/F strategy demonstrated its effectiveness w.r.t. the single-sensor JPDA-UKF, in terms of increased ToT (about 6% w.r.t. Palmaria and 21% w.r.t. San Rossore on average) and reduced RMSE (about 200 m and 0.5 m/s for position and velocity estimates). The aforementioned results demonstrated that radar network systems can take advantage from more advanced signal processing techniques and from aspect diversity for providing additional information on the maritime picture with no further costs on the system setup.

5.8 Summary

Nowadays, operative surveillance scenarios continuously bring an ever-increasing number of challenges, thus entailing the absolute necessity of exploiting as much as information as possible. In the face of this increasingly complex scenario and dense electromagnetic environments, radar networks have turned out to be an absolute necessity in order to achieve accurate and comprehensive information. In this chapter, the properties, the characteristics, and the main challenges of radar networks and data fusion strategies have been presented and analyzed.

First, an overview of the main concepts, benefits, and possible applications of radar networks were investigated. Second, two main radar network categorizations were discussed. With the first categorization it is possible to distinguish among monostatic networks, when only monostatic radar are present, bistatic networks, when there is only one transmitter and all others receivers sensors (thus forming bistatic radar pairs), and a combination of monostatic and bistatic (multistatic case) radars. The second possible categorization of radar networks allows to distinguish between centralized and decentralized processing approaches. In particular, the advantages and disadvantages of centralized and decentralized approaches were considered, with particular attention to system performance and additional computational complexity trade-offs between the two signal processing approaches. The centralized approach is based on the use of a central processing unit, which collects all the raw information and implements a coherent combination of the received signals. However, this approach assumes a perfect synchronization among the network sensor nodes and high reliable communication links. Unfortunately, these assumptions are very difficult to be met. On the other hand, in the decentralized approach, each radar system pre-processes its own data before transmitting them to the central processing unit, thus reducing the volume of information to be exchanged. Even though the decentralized approach is considered suboptimum, due to its intrinsic robustness against node failures, it is the most common used approach in operative netted radar systems.

After the description of the benefits and the higher capabilities of the radar network respect to monostatic radar system, in this chapter have been then discussed the main issues for the realizations of network radar system: the synchronization problem and the data fusion information methods. The synchronization between the radar network nodes is crucial to guarantee the operational use of a radar network and to perform both centralized and decentralized processing approach. Therefore, the impact of the synchronization on radar network system performance has been evaluated and the principal causes of radar network synchronization loss have been investigated in order to design the effective countermeasures for counteracting the mis-synchronization among the network nodes.

After the description of the benefits and the higher capabilities of the radar network respect to monostatic radar system, in this chapter have been then discussed the main issues for the realization of a radar network: the synchronization problem and the data fusion methods. Synchronization between the radar network nodes is crucial to guarantee the operational use of a radar network and to allow the use of either centralized or decentralized processing approaches. As a result, the impact of synchronization on radar network performance was evaluated and the principal causes of synchronization loss were investigated in order to better design the effective countermeasures.

An overview of the data fusion methods was considered for different radar network architectures, with particular emphasis on centralized and decentralized network paradigm, and, to hybrid options too. The main characteristics of the most common data fusion methods and a review of the state of the art for the MTT problem were then provided.

Finally, a recent experimentation conducted with two operative HFSW radars for maritime surveillance was described. In order to assess the performance of the radar network, a methodology for validating the detection and tracking capabilities, together with the description of performance metrics, was proposed and motivated. These metrics (e.g., the ToT and FAR) were evaluated for the fused radar system, then, they were compared with those obtained by using single-operating radars. The comparison analysis demonstrated that the angular diversity provided by the radar network allows to obtain better results (i.e., increased ToT and reduced RMSE of the target state vector estimate), tracking thus proving the effectiveness and usefulness of radar networks in the maritime surveillance framework.

References

[1] C. J. Baker, “An introduction to multistatic radar,” NATO-SET 136 Lecture Series Multistaic Surveillance and Reconnaissance: Sensor, Signal and Data Fusion, 2009.

[2] C. Baker and A. Hume, “Netted radar sensing,” *IEEE Aerospace and Electronic Systems Magazine*, vol. 18, no. 2, pp. 3–6, 2003.

[3] J. Li and P. Stoica, *MIMO radar signal processing.*, Wiley Online Library, 2009.

[4] H. Griffiths, “Bistatic and multistatic radar,” *University College London Dept. Electronic and Electrical Engineering*, 2004.

[5] C. J. Baker and H. Griffiths, “Bistatic and multistatic radar sensors for homeland security,” *Advances in Sensing with Security Applications*, pp. 1–22, 2006.

[6] V. Chernyak, *Fundamentals of Multisite Radar Systems*, London: Routledge, 1998. https://doi.org/10.1201/9780203755228.

[7] E. Hanle, “Survey of bistatic and multistatic radar,” *Communications, Radar and Signal Processing, IEE Proceedings F*, vol. 133, no. 7, pp. 587–595, 1986.

[8] Papoutsis, C. Baker, and H. Griffiths, “Fundamental performance limitations of radar networks,” in *1st EMRS DTC Tech. Conf.*, Edinburgh, UK, 2004.

[9] T. Johnsen, “Time and frequency synchronization in multistatic radar. Consequences to usage of GPS disciplined references with and without GPS signals,” in *Radar Conference, 2002. Proceedings of the IEEE*, Long Beach, CA: USA, 2002, pp. 141–147.

[10] C. Yougguang, L. Xicheng, Q. Hua, and J. Xiaojun, “On study of the application of ATM switches in netted-radar systems,” in *Aerospace and Electronics Conference, 1997. NAECON 1997, Proceedings of the IEEE 1997 National*, vol. 2, IEEE, Dayton, OH: USA, 1997, pp. 970–974.

[11] V. S. Chernyak, “Multisite radar systems with information fusion: A technology of XXI century.” Available at: https://www.researchgate.net/profile/Victor_Chernyak/publication/255664974_Multisite_Radar_Systems_with_Information_Fusion_A_Technology_of_XXI_Century/links/0f3175368846840370000000/Multisite-Radar-Systems-with-Information-Fusion-A-Technology-of-XXI-Century.pdf

[12] H. Griffiths, “Multistatic, mimo and networked radar: the future of radar sensors?” in *Radar Conference (EuRAD), 2010 European*, IEEE, Paris, France, 2010, pp. 81–84.

[13] C. J. Baker, H. Griffiths, and M. Vespe, “Multi-perspective imaging and image interpretation,” in *Imaging for Detection and Identification*, Dordrecht, Springer, 2007, pp. 1–28.

[14] M. Swift, J. Riley, S. Lourey, and L. Booth, “An overview of the multistatic sonar program in Australia,” in *Signal Processing and Its Applications, 1999. ISSPA'99. Proceedings of the Fifth International Symposium on*, vol.1, IEEE, Brisbane, Queensland, Australia, 1999, pp. 321–324.

[15] P. Braca, P. Willett, K. D. LePage, S. Marano, and V. Matta, "Bayesian tracking in underwater wireless sensor networks with port-starboard ambiguity." *IEEE Trans. Sign. Process.*, vol. 62, no. 7, pp. 1864–1878, 2014.

[16] G. Ferri, A. Tesei, P. Braca, *et al.*, "Cooperative robotic networks for underwater surveillance: an overview," *IET Radar, Sonar & Navigation*, 2017.

[17] T. A. Seliga and F. J. Coyne, "Multistatic radar as a means of dealing with the detection of multipath false targets by airport surface detection equipment radars," in *Radar Conference, 2003. Proceedings of the 2003 IEEE*, Huntsville, AL: USA 2003, pp. 329–336.

[18] C. Fischer, M. Younis, and W. Wiesbeck, "Multistatic GPR data acquisition and imaging," in *Geoscience and Remote Sensing Symposium, 2002. IGARSS'02. 2002 IEEE International*, vol. 1, IEEE, Toronto, Ontario, Canada, 2002, pp. 328–330.

[19] H. Groll, J. Detlefsen, and U. Siart, "Multi sensor systems at mm-wave range for automotive applications," in *Radar, 2001 CIE International Conference on, Proceedings*, IEEE, Beijing, China, 2001, pp. 150–153.

[20] W. G. Bath, "Tradeoffs in radar networking," *IET Conference Proceedings*, Edinburgh, UK, pp. 26–30(4), January 2002.

[21] H. Griffiths, C. Baker, P. Sammartino, and M. Rangaswamy, "MIMO as distributed radar system," *MIMO Radar Signal Processing*, pp. 319–363, 2008.

[22] P. F. Sammartino, "A comparison of processing approaches for distributed radar sensing," Ph.D. dissertation, UCL (University College London), 2009.

[23] P. Sammartino, C. Baker, and M. Rangaswamy, "Moving target localization with multistatic radar systems," in *Radar Conference, 2008. RADAR'08. IEEE*, Rome, Italy, 2008, pp. 1–6.

[24] Y. Yang and R. S. Blum, "Phase synchronization for coherent mimo radar: algorithms and their analysis," *IEEE Trans. Sign. Process.*, vol. 59, no. 11, pp. 5538–5557, 2011.

[25] S. M. Hurley, M. Tummala, T. Walker, and P. E. Pace, "Impact of synchronization on signal-to-noise ratio in a distributed radar system," in *System of Systems Engineering, 2006 IEEE/SMC International Conference on*, IEEE, Taipei, Taiwan, 2006, pp. 5.

[26] B. Sundararaman, U. Buy, and A. D. Kshemkalyani, "Clock synchronization for wireless sensor networks: a survey," *Ad Hoc Netw.*, vol. 3, no. 3, pp. 281–323, 2005.

[27] Q. Li and D. Rus, "Global clock synchronization in sensor networks," *IEEE Trans. Comp.*, vol. 55, no. 2, pp. 214–226, 2006.

[28] D. L. Hall and J. Llinas, "An introduction to multisensor data fusion," *Proc. IEEE*, vol. 85, no. 1, pp. 6–23, 1997.

[29] D. Hall and J. Llinas, *Multisensor Data Fusion*, Boca Raton, FL, USA, CRC Press, 2001.

[30] E. Blasch, É. Bossé, and D. A. Lambert, *High-Level Information Fusion Management and Systems Design*, Boston, London, Artech House, 2012.

[31] T. Hughes, "Sensor fusion in a military avionics environment," *Measur. Contr.*, vol. 22, no. 7, pp. 203–205, 1989.

[32] T. Neumann, "Multisensor data fusion in the decision process on the bridge of the vessel," *Int. J. Marine Navig. Saf. Sea Transport.*, vol. 2, no. 1, pp. 85–89, 2008.

[33] E. Blasch, "Level 5 (user refinement) issues supporting information fusion management," in *Information Fusion, 2006 9th International Conference on*, IEEE, Florence, Italy, 2006, pp. 1–8.

[34] N. Xiong and P. Svensson, "Multi-sensor management for information fusion: issues and approaches," *Inform. Fus.*, vol. 3, no. 2, pp. 163–186, 2002.

[35] H. Durrant-Whyte, M. Stevens, and E. Nettleton, "Data fusion in decentralised sensing networks," in *4th International Conference on Information Fusion*, Montréal, Quebec, Canada, 2001, pp. 302–307.

[36] C.-Y. Chong, S. Mori, and K.-C. Chang, "Distributed multitarget multisensor tracking," *Multitarg. Multisens. Track. Adv. Appl.*, vol. 1, pp. 247–295, 1990.

[37] P. Greenway and R. H. Deaves, "Sensor management using the decentralized Kalman filter," in *Photonics for Industrial Applications.*, International Society for Optics and Photonics, Boston, MA, USA, 1994, pp. 216–225.

[38] G. W. Ng and K. H. Ng, "Sensor management – what, why and how," *Inform. Fus.*, vol. 1, no. 2, pp. 67–75, 2000.

[39] S. J. Julier and J. K. Uhlmann, "New extension of the Kalman filter to nonlinear systems," in *AeroSense'97.*, International Society for Optics and Photonics, Orlando, FL, USA, 1997, pp. 182–193.

[40] F. Castanedo, "A review of data fusion techniques," *Scient. World J.*, vol. 2013, pp. 1–19, 2013.

[41] H. W. Sorenson, "Least-squares estimation: from Gauss to Kalman," *IEEE Spectr.*, vol. 7, no. 7, pp. 63–68, 1970.

[42] Y. Bar-Shalom, P. K. Willett, and X. Tian, *Tracking and data fusion*, Storrs, CT, USA, YBS Publishing, 2011.

[43] R. P. Mahler, "A unified foundation for data fusion," *SPIE MILESTONE SERIES MS*, vol. 124, pp. 325–345, 1996.

[44] Poore and N. Rijavec, "Partitioning multiple data sets via multidimensional assignments and Lagrangian relaxation. Dimacs series in discrete mathematics and theoretical computer science," *American Mathematical Society, Providence, RI*, 1995.

[45] R. R. Yager, "On the Dempster-Shafer framework and new combination rules," *Inform. Sci.*, vol. 41, no. 2, pp. 93–137, 1987.

[46] K. Sentz and S. Ferson, *Combination of evidence in Dempster-Shafer theory*, Sandia National Laboratories Albuquerque, 2002, vol. 4015.

[47] P. Dempster, "A generalization of Bayesian inference." *Classic Works of the Dempster-Shafer Theory of Belief Functions*, vol. 219, pp. 73–104, 2008.

[48] P. Jackson, *Introduction to expert systems*. Boston, MA, USA: Addison-Wesley Longman Publishing Co., Inc., 1998.

[49] B. Hayes-Roth, "A blackboard architecture for control," *Artif. Intell.*, vol. 26, no. 3, pp. 251–321, 1985.

[50] G. J. Klir and Bo Yuan, *Fuzzy Sets and Fuzzy Logic Theory and Applications*. Vol. 4, Prentice Hall, Upper Saddle River, NJ, USA, p. 574, 1995.

[51] J. K. Uhlmann, "Algorithms for multiple-target tracking," *Am. Scient.*, vol. 80, no. 2, pp. 128–141, 1992.

[52] C. Rasmussen and G. D. Hager, "Probabilistic data association methods for tracking complex visual objects," *IEEE Trans. Pattern Anal. Mach. Intell.*, vol. 23, no. 6, pp. 560–576, 2001.

[53] R. J. Adrian, "Particle-imaging techniques for experimental fluid mechanics," *Ann. Rev. Fluid Mech.*, vol. 23, no. 1, pp. 261–304, 1991.

[54] Genovesio, T. Liedl, V. Emiliani, W. J. Parak, M. Coppey-Moisan, and J.-C. Olivo-Marin, "Multiple particle tracking in 3-d+ t microscopy: method and application to the tracking of endocytosed quantum dots," *IEEE Trans. Image Process.*, vol. 15, no. 5, pp. 1062–1070, 2006.

[55] S. Thrun, W. Burgard, and D. Fox, *Probabilistic Robotics*. Cambridge, MA, USA, MIT Press, 2005.

[56] J. S. Mullane, B.-N. Vo, M. D. Adams, and B.-T. Vo, *Random Finite Sets for Robot Mapping & SLAM: New Concepts in Autonomous Robotic Map Representations*. Berlin, Heidelberg, Springer Science & Business Media, 2011, vol. 72.

[57] K. Witrisal, P. Meissner, E. Leitinger, *et al.*, "High-accuracy localization for assisted living: 5g systems will turn multipath channels from foe to friend," *IEEE Sign. Process. Mag.*, vol. 33, no. 2, pp. 59–70, 2016.

[58] E. Leitinger, F. Meyer, P. Meissner, K. Witrisal, and F. Hlawatsch, "Belief propagation based joint probabilistic data association for multipath-assisted indoor navigation and tracking," in *Localization and GNSS (ICL-GNSS), 2016 International Conference on*. IEEE, Barcelona, Spain, 2016, pp. 1–6.

[59] C. Urmson, J. Anhalt, D. Bagnell, *et al.*, "Autonomous driving in urban environments: Boss and the urban challenge," *J. Field Robot.*, vol. 25, no. 8, pp. 425–466, 2008.

[60] J. Levinson, J. Askeland, J. Becker, *et al.*, "Towards fully autonomous driving: Systems and algorithms," in *Intelligent Vehicles Symposium (IV), 2011 IEEE*. IEEE, Baden-Baden, Germany, 2011, pp. 163–168.

[61] S. Reuter and K. Dietmayer, "Pedestrian tracking using random finite sets," in *Information Fusion (FUSION), 2011 Proceedings of the 14th International Conference on*, *IEEE*, Chicago, IL, USA, 2011, pp. 1–8.

[62] R. W. Sittler, "An optimal data association problem in surveillance theory," *IEEE Trans. Milit. Electron.*, vol. 8, no. 2, pp. 125–139, 1964.

[63] T. Fortmann, Y. Bar-Shalom, and M. Scheffe, "Sonar tracking of multiple targets using joint probabilistic data association," *IEEE J. Ocean. Eng.*, vol. 8, no. 3, pp. 173–184, 1983.

[64] R. Singer, R. Sea, and K. Housewright, "Derivation and evaluation of improved tracking filter for use in dense multitarget environments," *IEEE Trans. Inf. Theory*, vol. 20, no. 4, pp. 423–432, 1974.

[65] Y. Bar-Shalom and E. Tse, "Tracking in a cluttered environment with probabilistic data association," *Automatica*, vol. 11, no. 5, pp. 451–460, 1975.

[66] Y. Bar-Shalom, "Tracking methods in a multitarget environment," *IEEE Trans. Autom. Control*, vol. 23, no. 4, pp. 618–626, 1978.

[67] D. B. Reid, "An algorithm for tracking multiple targets," *IEEE Trans. Autom. Control*, vol. 24, no. 6, pp. 843–854, 1979.

[68] R. Niu, P. Willett, and Y. Bar-Shalom, "Matrix CRLB scaling due to measurements of uncertain origin," *IEEE Trans. Sign. Process.*, vol. 49, no. 7, pp. 1325–1335, 2001.

[69] Y. Bar-Shalom, "Tracking methods in a multitarget environment," *IEEE Trans. Autom. Contr.*, vol. 23, no. 4, pp. 618–626, 1978.

[70] T. Kirubarajan and Y. Bar-Shalom, "Probabilistic data association techniques for target tracking in clutter," *Proc. IEEE*, vol. 92, no. 3, pp. 536–557, 2004.

[71] S. S. Blackman, "Multiple hypothesis tracking for multiple target tracking," *IEEE Trans. Aerosp. Electron. Syst.*, vol. 19, pp. 5–18, 2004.

[72] E. Mazor, A. Averbuch, Y. Bar-Shalom, and J. Dayan, "Interacting multiple model methods in target tracking: a survey," *IEEE Trans. Aerosp. Electron. Syst.*, vol. 34, no. 1, pp. 103–123, 1998.

[73] T. Kirubarajan, Y. Bar-Shalom, K. R. Pattipati, and I. Kadar, "Ground target tracking with variable structure IMM estimator," *IEEE Trans. Aerosp. Electron. Syst.*, vol. 36, no. 1, pp. 26–46, 2000.

[74] K.-C. Chang and Y. Bar-Shalom, "Joint probabilistic data association for multitarget tracking with possibly unresolved measurements and maneuvers," *IEEE Trans. Autom. Control*, vol. 29, no. 7, pp. 585–594, 1984.

[75] J.W. Koch, "Bayesian approach to extended object and cluster tracking using random matrices," *IEEE Aerosp. Electron. Syst.*, vol. 44, no. 3, pp. 1042–1059, 2008.

[76] G. Pulford, "Taxonomy of multiple target tracking methods," *IEE Proceed. –Rad., Sonar Navig.*, vol. 152, no. 5, pp. 291–304, 2005.

[77] Y. Bar-Shalom and X.-R. Li, "Multitarget-multisensor tracking: principles and techniques," Storrs, CT: University of Connecticut, 1995.

[78] R. P. Mahler, "Multitarget Bayes filtering via first-order multitarget moments," *IEEE Trans. Aerosp. Electron. Syst.*, vol. 39, no. 4, pp. 1152–1178, 2003.

[79] B.-N. Vo, S. Singh, and A. Doucet, "Sequential Monte Carlo methods for multitarget filtering with random finite sets," *IEEE Trans. Aerosp. Electron. Syst.*, vol. 41, no. 4, pp. 1224–1245, 2005.

[80] R. P. Mahler, *Statistical Multisource-Multitarget Information Fusion*, Norwood, MA, USA, Artech House, 2007.

[81] P. Braca, S. Marano, V. Matta, and P. Willett, "Asymptotic efficiency of the PhD in multitarget/multisensor estimation," *IEEE J. Select. Topics Sig. Process.*, vol. 7, no. 3, pp. 553–564, 2013.

[82] B.-T. Vo, B.-N. Vo, and A. Cantoni, "Analytic implementations of the cardinalized probability hypothesis density filter," *IEEE Trans. Sign. Process.*, vol. 55, no. 7, pp. 3553–3567, 2007.

[83] B.-T. Vo, B.-N. Vo, and A. Cantoni, "The cardinality balanced multi-target multi-Bernoulli filter and its implementations," *IEEE Trans. Sign. Process.*, vol. 57, no. 2, pp. 409–423, 2009.

[84] B.-T. Vo and B.-N. Vo, "Labeled random finite sets and multi-object conjugate priors," *IEEE Trans. Sign. Process.*, vol. 61, no. 13, pp. 3460–3475, 2013.

[85] B.-N. Vo, B.-T. Vo, and D. Phung, "Labeled random finite sets and the Bayes multi-target tracking filter," *IEEE Trans. Sign. Process.*, vol. 62, no. 24, pp. 6554–6567, 2014.

[86] O. Cappé, S. J. Godsill, and E. Moulines, "An overview of existing methods and recent advances in sequential Monte Carlo," *Proceed. IEEE*, vol. 95, no. 5, pp. 899–924, 2007.

[87] M. R. Morelande, C. M. Kreucher, and K. Kastella, "A Bayesian approach to multiple target detection and tracking," *IEEE Trans. Sign. Process.*, vol. 55, no. 5, pp. 1589–1604, 2007.

[88] F. R. Kschischang, B. J. Frey, and H.-A. Loeliger, "Factor graphs and the sum-product algorithm," *IEEE Trans. Inf. Theory*, vol. 47, no. 2, pp. 498–519, 2001.

[89] J. S. Yedidia, W. T. Freeman, and Y. Weiss, "Constructing free-energy approximations and generalized belief propagation algorithms," *IEEE Trans. Inf. Theory*, vol. 51, no. 7, pp. 2282–2312, 2005.

[90] P. Horridge and S. Maskell, "Real-time tracking of hundreds of targets with efficient exact JPDAF implementation," in *Information Fusion, 2006 9th International Conference on*, IEEE, Florence, Italy, 2006, pp. 1–8.

[91] Z. Chena, L. Chen, M. Cetin, and A. S. Willsky, "An efficient message passing algorithm for multi-target tracking," in *Information Fusion, 2009. FUSION'09. 12th International Conference on, IEEE*, Seattle, WA, USA, 2009, pp. 826–833.

[92] J. Williams and R. Lau, "Approximate evaluation of marginal association probabilities with belief propagation," *IEEE Tran. Aerosp. Electron. Syst.*, vol. 50, no. 4, pp. 2942–2959, 2014.

[93] J. L. Williams, "Marginal multi-Bernoulli filters: RFs derivation of MHT, JIPDA, and association-based member," *IEEE Trans. Aerosp. Electron. Syst.*, vol. 51, no. 3, pp. 1664–1687, 2015.

[94] F. Meyer, P. Braca, P. Willett, and F. Hlawatsch, "Scalable multitarget tracking using multiple sensors: a belief propagation approach," in *Information Fusion (Fusion), 2015 18th International Conference on*, IEEE, Washington, DC, USA, 2015, pp. 1778–1785.

[95] F. Meyer, P. Braca, P. Willett, and F. Hlawatsch, "Tracking an unknown number of targets using multiple sensors: a belief propagation method," in *Information Fusion (FUSION), 2016 19th International Conference on*, IEEE, Heidelberg, Germany, 2016, pp. 719–726.

[96] F. Meyer, P. Braca, P. Willett, and F. Hlawatsch, "A scalable algorithm for tracking an unknown number of targets using multiple sensors," *arXiv preprint arXiv:1607.07647*, 2016.

[97] P. Braca, S. Maresca, R. Grasso, K. Bryan, and J. Horstmann, "Maritime surveillance with multiple over-the-horizon HFSW radars: an overview of recent experimentation," *IEEE Aerosp. Electron. Syst. Mag.*, vol. 30, no. 12, pp. 4–18, 2015.

[98] K. Granström, A. Natale, P. Braca, G. Ludeno, and F. Serafino, "Gamma Gaussian inverse Wishart probability hypothesis density for extended target tracking using x-band marine radar data," *IEEE Trans. Geosci. Rem. Sens.*, vol. 53, no. 12, pp. 6617–6631, 2015.

[99] R. Goldhahn, P. Braca, K. D. LePage, P. Willett, S. Marano, and V. Matta, "Environmentally sensitive particle filter tracking in multistatic AUV networks with port-starboard ambiguity," in *Acoustics, Speech and Signal Processing (ICASSP), 2014 IEEE International Conference on*, IEEE, Florence, Italy, 2014, pp. 1458–1462.

[100] P. Braca, R. Goldhahn, K. D. LePage, S. Marano, V. Matta, and P. Willett, "Cognitive multistatic AUV networks," in *Information Fusion (FUSION), 2014 17th International Conference on*, IEEE, Salamanca, Spain, 2014, pp. 1–7.

[101] D. Schuhmacher, B.-T. Vo, and B.-N. Vo, "A consistent metric for performance evaluation of multi-object filters," *IEEE Trans. Sign. Process.*, vol. 56, no. 8, pp. 3447–3457, 2008.

[102] K.-W. Gurgel and T. Schlick, "HF radar wave measurements in the presence of ship echoes – problems and solutions," in *Oceans 2005-Europe*, vol. 2.IEEE, 2005, pp. 937–941.

[103] S. Maresca, P. Braca, J. Horstmann, and R. Grasso, "Maritime surveillance using multiple high-frequency surface-wave radars," *IEEE Trans. Geosci. Rem. Sens.*, vol. 52, no. 8, pp. 5056–5071, 2014.

[104] Dzvonkovskaya, K.-W. Gurgel, H. Rohling, and T. Schlick, "Low power high frequency surface wave radar application for ship detection and tracking," in *Radar, 2008 International Conference on, IEEE*, Adelaide, SA, Australia, 2008, pp. 627–632.

Chapter 6

Photonics in radar networks

Sergio Pinna[1], Salvatore Maresca[2], Francesco Laghezza[3], Leonardo Lembo[2,4] and Paolo Ghelfi[5]

6.1 Chapter organization and key points

In this chapter, we discuss the advantages of exploiting photonic techniques in radar networks.

More in detail, we first review the specific requirements of radar networks, focusing in particular on the coherence requested to the radar signals at the different network nodes. Then, we discuss the potentials of using photonics for distributing synchronization signals through optical fibers in distributed radar networks. Finally, we concentrate on the most performing radar network architecture, namely the centralized radar network, to report very recent experimental results enabled by photonics, and analyze the potential of a multi-input–multi-output (MIMO) processing approach in a radar network with widely separated antennas.

As will be clear from the discussions reported herewith, photonics is expected to become fundamental for further improving the performance of radar systems by allowing the implementations of wide radar networks that can gather more information from the observed scenario thanks to the geometrical diversity and sustained by the easiness of the synchronization methods.

Although the photonics-based solutions described here are still in an embryonal advancement status, the way for a next quantum leap in radar networks architectures seems clearly paved.

6.2 Coherence and synchronization in radar networks

The developments in radar engineering have mainly focused so far on monostatic or bistatic radar configurations, which exploit only one transmitter and one receiver

[1]Department of Electrical and Computer Engineering (ECE), University of California at Santa Barbara (UCSB), USA
[2]Institute of Technologies for Communication, Information and Perception (TeCIP), Sant'Anna School of Advanced Studies, Italy
[3]NXP Semiconductors, The Netherlands
[4]Italian Navy, Naval Research Center (CSSN), Italy
[5]National Laboratory of Photonic Networks and Technology (PNTLab), Consorzio Nazionale Interuniversitario per le Telecomunicazioni (CNIT), Italy

that can be co-located or widely separated. However, these radar architectures are not always sufficient to face the new emerging and challenging operative scenarios, where the number and variety of threats can be almost unlimited [1,2] (see also Section 5.2). For this reason, the concept of multistatic radars (also called multisensor or multisite radars) has been introduced [3–6].

Multistatic radars, or radar networks as they are often called [7], employ several spatially distributed transmitting and receiving "nodes" (i.e., antenna sites) allowing the observation of a scenario from multiple viewpoints. This makes them particularly suited for surveillance applications since they permit an improved angular resolution and therefore an ability to separate closely spaced multiple targets [4,8,9]. Moreover, the multiple viewpoints allow detecting the signal scattered by complex targets characterized by multiperspective radar cross section (RCS) (i.e., stealth targets) [10,11]. In addition, multistatic radars can handle slow-moving targets by exploiting Doppler estimates from multiple directions [12] and feature a highly accurate estimate of target position [13,14].

Radar networks can be implemented in the form of several different architectures, each with peculiar strengths and weaknesses. As will be explained in the following, photonics is promising a significant simplification in the development of the most performing ones, in particular because it enables a high coherence and a straightforward synchronization of the network nodes.

Here, we review the possible classifications of radar networks and recall the features and issues of the node synchronization methods.

6.2.1 Classification of radar networks

6.2.1.1 Non-coherent versus coherent radar networks

The large majority of modern monostatic or bistatic radars are coherent systems, meaning that they can measure the phase and frequency shift of the received signal with respect to the reference signal generated by their internal highly stable local oscillator. Any relative change in the received signal phase or frequency can therefore be attributed to the target characteristics, mainly its range and velocity [15].

In radar networks, the concept of coherence must be extended to the entire network. Depending on the implementation, we can distinguish between noncoherent and coherent radar networks [7].

In noncoherent networks (Figure 6.1(a)), each of the radar nodes acts as an autonomous system, independently calculating the target parameters such as distance and velocity. If N nodes compose the radar network, the independent nodes provide N different observations of the same scene. These observations are then gathered by a "command & control" center that combines the received data and extracts an improved description of the scene thanks to the geometrical disposition of the nodes, which allows looking at the target from different viewpoints.

In a noncoherent network, although the nodes are independent, they need to share a common sense of space and time to avoid introducing errors in the description of the scene [7]. Therefore, the command & control center needs to know the positions of the nodes, and the gathered data needs to include a time tag, which should refer to a common timing system. The precision required in the time synchronization is in the order of a fraction of the compressed radar pulse (typically, down to few nanoseconds) [1]. This can be given, for example, by the timing signal from Global Navigation Satellite Systems (GNSS) such as the Global Positioning System (GPS). More details on time synchronization will be given in the following.

Since the data from the different nodes is not coherent, the processing in this kind of radar networks is usually simple, limited to the fusion of the target detections or tracks from each node. Moreover, as the nodes and the command & control center share data that are already processed (e.g., detections or tracks with related position and velocity), the necessary communication links between them must guarantee a fairly narrow bandwidth (in the order of few MB/s).

The case study reported in Section 5.7 is a significant example of a noncoherent radar network.

In a coherent radar network with N transmitting and receiving nodes, each node can receive the echo of the signal transmitted by any of the N nodes (Figure 6.1(b)). Therefore, this monitoring system can provide up to N^2 different observations of the same scenario, guaranteeing an improved robustness against target parameters fluctuations, background variations, and interference. (This discussion can be easily extended to systems with M transmitters and N receivers, with $M \neq N$, and $N \times M$ different observations.)

In order to allow any receiver detecting the signal from any transmitter, all the transmitters and receivers must share the same carrier frequency, and the phase relations between all the nodes must be known and constant. Moreover, a precise common timing must also be guaranteed (in fact, the phase synchronization does not imply the synchronization of the radar pulses). The precision required in the synchronization is therefore in the order of a fraction of the carrier period (i.e., down to few ps in the case of a X-band radar) [1]. More details on phase synchronization will be given in the following. Of course, as for the noncoherent case, the exact position of the network nodes must be known as well.

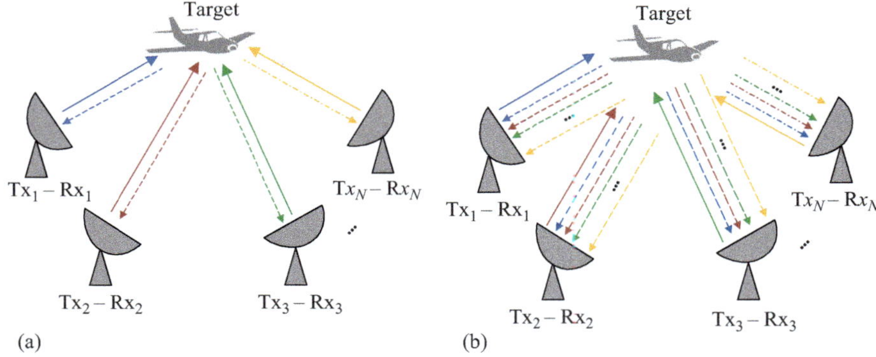

Figure 6.1 (a) Noncoherent and (b) coherent radar networks

It is fundamental to underline that in a coherent radar network, besides the improvement in the number of viewpoints described above coming from the geometrical diversity, the different observations can be all processed coherently, thus increasing the signal-to-noise ratio (SNR) of the overall detected echo. Assuming that all the nodes in the network are identical and perfectly synchronized, and that the target is an isotropic radiator [1,7], the general range equation of the radar network can be expressed as:

$$SNR = \sum_{i=1}^{N}\sum_{j=1}^{N} \frac{P_T G_T G_R \tau \sigma \lambda^2}{(4\pi)^3 k T_0 F R_{T_i}^2 R_{R_j}^2 L}, \tag{6.1}$$

where P_T is the transmitted power, G_T and G_R are the transmitter/receiver antenna gains, τ is the pulse length (supposing to transmit unmodulated pulses), σ is the target RCS, λ is the carrier wavelength, k is the Boltzmann's constant, T_0 is the system noise temperature, F is the noise figure, R_{Ti} and R_{Rj} are the range from transmitter i to the target and from the target to receiver j, respectively, while L describes the overall system losses. As can be seen from this equation, under the assumptions described above and in the case of a perfectly symmetric geometry, the SNR of the overall detected echo can be N^2 times larger than the SNR of the signal detected by the single node, that is, the radar network ensures an SNR gain as high as N^2.

The powerful coherent processing described above can be implemented only by merging the raw measurements coming from each node. Due to the large data rate of the raw information, this kind of data fusion is computationally heavy. Moreover, the communication links between the network nodes must be broadband enough to allow transferring the raw data as a continuous stream (which depends on the signal bandwidth and can reach challenging values as high as few GB/s, with several tens of MB/s being common figures).

Summarizing, coherent radar networks are significantly more performing than noncoherent ones, since they allow

- a larger number of geometrically different observations (N^2 instead of N, in the N-node configuration considered above) and
- a coherent processing of raw data that can increase significantly the SNR of the fused signal (with a SNR gain up to N^2, with respect to the SNR of the signal detected at a single node).

On the other hand, coherent networks require a significantly more complex synchronization (as will be discussed below) and pose challenges on the communication links between nodes to allow transferring raw data.

As will be described in the rest of this chapter, the photonics-based signal distribution over optical fiber could play a key role for pushing the development of coherent radar networks.

Finally, it is worth noting here that the observed scenario has an influence on the coherence of its observation. In fact, the coherence in the radar network allows extending the integration time, but this makes sense only if the observed scenario is stationary during the observation (i.e., during the coherent integration time [CIT]). This note suggests that the requirement on coherence posed on a radar network must take into due account the stationarity of the scenario as well.

6.2.1.2 Distributed versus centralized radar networks

A further classification of radar networks can be based on where signal generation and processing are made. Thus, radar networks can be divided into centralized and distributed architectures [3,4]. These network types are shown in Figure 6.2. Hierarchical/hybrid architectures exist as well, as stated in Chapter 5, but for simplicity they will not be addressed in this chapter.

In distributed radar networks, each remote node generates, transmits, receives, and preprocesses the radar signal (Figure 6.2(a)). Since the nodes must share at least a common timing, as highlighted above, one of the nodes acts as master node and distributes a synchronization clock. Moreover, the master node receives the pre-processed signals from the other nodes for further processing and for data fusion [6].

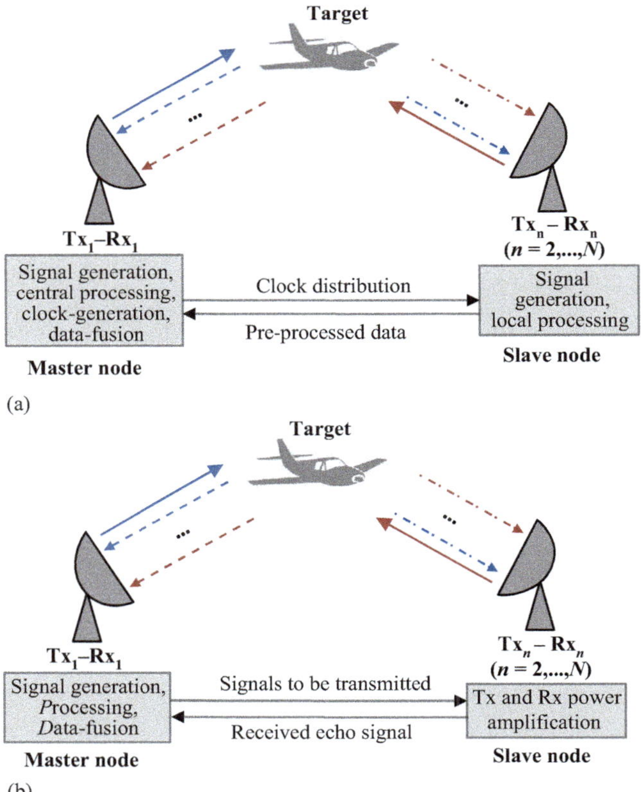

Figure 6.2 (a) Distributed network architecture and (b) centralized network architecture

Table 6.1 Synchronization requirements for distributed and centralized radar networks

	Distributed	**Centralized**
Noncoherent	Timing synchronization required	–
Coherent	Time and phase synchronization required, broadband link required	Time and phase synchronization is guaranteed

A communication link is therefore necessary between the master node and the slave nodes, capable of distributing the clock signal and of collecting the preprocessed data.

In centralized radar networks, the master node performs the entire signal generation and processing, and acts as the command & control center. The slave nodes instead only operate signal amplification and, in some cases, signal frequency up- and down-conversion.

The bi-directional links between the master and slave nodes must provide sufficient bandwidth to transfer the transmitted signal and the received echoes without introducing significant distortions.

In such an architecture, since all the signals are generated and processed by the same unit, signal synchronization is highly simplified, and the network is intrinsically coherent.

As we have seen above, the different network architectures impose different requirements in terms of synchronization and link bandwidth, as summarized in Table 6.1. In the following, we analyze these aspects in more detail.

6.2.2 Synchronization in radar networks

If we consider a single radar system, the synchronization among the radar constituent blocks represents one of the most crucial issues to cope with, involving the transmitter and the receiver, as well as the target. A less-than-perfect synchronization between these components can affect the generation of radar signals, the internal communications and the electronic devices. These can have a negative effect on transmitter timing, receiver protection timing, analog-to-digital-converter (ADC) sample timing, and pulse repetition frequency.

When considering radar networks, in addition to the aforementioned issues, adequate synchronization in terms of time, frequency, and phase is required among the network nodes to maximize the final SNR [16–18]. In fact, the incorrect combination of pulses due to node synchronization errors reduces the SNR, thus deteriorating the overall radar network performance.

In centralized radar networks, as discussed above, the timing is dictated by the master node and by its central processor unit (see Table 6.1). Therefore, the entire network is intrinsically synchronized.

In distributed radar networks, each node has its own internal processor clock and a synchronization is required [19].

In noncoherent distributed networks, the synchronization involves only a common timing.

Instead, in coherent distributed networks the synchronization also requires a common phase for each of the nodes. Two main approaches to phase synchronization exist: the master–slave and the peer-to-peer strategies [20]. In the master–slave paradigm, one node is identified as the master, and the slave nodes consider the master local clock as the phase reference, synchronizing and locking their local clock to it. In the peer-to-peer paradigm, each sensor can communicate directly with every other node in the network, thus eliminating the issues of a possible master node failure, but with a significantly complex implementation [20]. Both the approaches require to share the clock signals through a direct link (e.g., coaxial cable, optical fiber, or wireless link), which can be easy for closely spaced nodes, or technically challenging if the nodes are distant [21].

Some details of these two approaches can be found in Section 5.4.

Before discussing the advantages of photonics in radar networks, it is worth describing briefly the issues on synchronization in distributed radar networks based on standard microwave technologies.

6.2.2.1 Time synchronization

The simplest network synchronization case is the time synchronization required in noncoherent distributed radar networks (see Table 6.1). In these systems, the time synchronization is needed in order to properly evaluate the time of flight of radar pulses, which directly affects the precision of the range measurement.

A typical precision requirement for this time synchronization is a fraction (usually, one-tenth) of the transmitted compressed pulse width [1,16,22], that is, down to few nanoseconds, which is usually not a challenge. Moreover, it is not necessary to provide an absolute time reference to the network nodes, instead a relative synchronization of the pulse emission instant is sufficient, further relaxing the timing requirement.

If the radar network utilizes a stable pulse repetition interval (PRI), the time synchronization can be obtained by periodically synchronizing identical oscillators at each network node to correct their slow drifts. If the nodes are all within line of sight (LOS), this periodic time synchronization can be done directly by transmitting a clock signal from the master to the slave nodes via dedicated communication links. Otherwise, the oscillators need to be indirectly slaved to a second source, such as the GPS signals. In this last case, specific GPS disciplined oscillators can be used at each node [23]. These solutions allow a time synchronization over long periods with errors of few nanoseconds.

If the PRI of the radar system is modulated (e.g., staggered, jittered, or randomly modulated), the time synchronization needs to be direct, sending a clock signal through dedicated communication links to synchronize the oscillator of the slave nodes. The communication link is often a direct radio link. If no LOS exists between the nodes, a dedicated satellite link, or a scatter path [24], or even the tropospheric scattering [25] can be used.

The time synchronization poses a requirement on the stability of the oscillator in the radar nodes. The required clock stability between updates can be defined as $\Delta\tau/T_u$, where $\Delta\tau$ is the required timing accuracy and T_u is the interval between clock updates. On one hand, the update interval can be as short as the PRI, but on the other hand it can be much longer, depending on the specific network implementation. A typical value is given by the transmitter antenna scan period (e.g., in the case of rotating antenna). As an example, if the required timing accuracy $\Delta\tau$ is specified as 0.1 μs, and the clock update interval T_u is the antenna scan period of 10 s, the required clock stability is 10^{-8} (one part in 10^8 over 10 s) [1].

In all the time synchronization schemes described above, the implementation is straightforward [1], much like the initial synchronization process in communication systems. With the time synchronization established, the target range can be correctly calculated.

6.2.2.2 Phase synchronization

In coherent distributed radar networks, besides the time synchronization, the phase information of the signals at the radar nodes need to be shared as well (see Table 6.1) [16,18]. In fact, the phase instability affects the maximum CIT and consequently the system sensitivity, while for a multistatic Doppler or moving target indicator processing, the phase coherence between the nodes has to be established.

Although the feasibility of distributed radar systems was already demonstrated by experimental implementations in [22,26–28], the time and phase synchronization is still an open issue hindering the development of distributed radars [29–33]. This issue has been investigated in particular in satellite synthetic aperture radar (SAR) systems [21,29,34–39].

As for the time synchronization, the phase synchronization between nodes can be implemented indirectly or directly.

The indirect synchronization can be implemented exploiting satellite signals. Typically, the "1 pulse-per-second" (1 PPS) reference signal provided by the GNSS systems is used to synchronize high-precision oscillators at the network nodes.

The direct phase synchronization can be implemented via land lines or other communication links. If the network nodes are in LOS, a direct wireless link can also be used. An extension of direct path phase locking is the use of the direct path signal as a reference signal in a correlation processor [40].

The need for phase synchronization poses a significant challenge on the stability of the oscillators. In fact, the phase stability has to be guaranteed during the whole CIT T_{CIT}, and it must be lower than $\Delta\varphi/2\pi f T_{CIT}$, with $\Delta\varphi$ the allowable root mean square phase error (in radians), and f the transmitter frequency (in Hz) [1]. Considering as an example a ground-based bistatic radar in the S-band with a center frequency of 3 GHz, and a maximum phase deviation $\Delta\varphi = 4°$ (0.07 rad) over a typical CIT of 1 s, the required oscillator stability is $3.7 \cdot 10^{-12}$ (i.e., about 2.7 parts in 10^{11} over 1 s). This requirement can be fulfilled with a high-performing temperature-controlled crystal oscillator (TCXO), as reported in [23]. However, if we use a radar carrier in the Ka-band instead, with a frequency of 30 GHz, and we keep unvaried both $\Delta\varphi$ and T_{CIT}, we obtain a required stability of $3.7 \cdot 10^{-13}$, and the TXCO is no longer suitable.

Besides the phase noise associated to the master clock, additional noise can be introduced by the clock distribution network. Such additive phase noise can come from the group index variations in the network transmission lines, caused by random temperature variations or mechanical stress [41]. In the case of wireless links, noise comes from the multipath and fading effects. Several techniques have been studied to reduce the phase noise impairments introduced by the clock and signal distribution network [17,21].

In the following, we will discuss how photonics can be used to manage the synchronization and data transfer in the case of coherent networks, either distributed or centralized.

6.3 Photonics for synchronizing distributed radar networks

As we have discussed in the previous paragraph, the phase synchronization in distributed radar networks can be implemented via wired lines (typically, coaxial cable) or wireless links. However, coaxial cables are heavy, lossy, and expensive, and can be adopted only in radar networks with closely spaced nodes. On the other hand, the effectiveness of wireless links depends on the carrier frequency: at high frequencies, for example, in the millimeter-wave frequency range (30–300 GHz), the wireless links are limited by the strong atmospheric attenuation. Moreover, wireless links require a good LOS to be effective.

Therefore, optical fiber networks are becoming the most reliable mean of signal exchange among the radar nodes, exploiting the radio-over-fiber approach [41], due to the very low optical signal attenuation, the large bandwidth available, and the RF immunity.

In a distributed coherent radar network (see Figure 6.2(a)), a master clock source, situated in the master node, generates the clock signal to be shared as a reference for frequency and phase locking among all the remote nodes. The clock source can be either electrical or optical, but in both the proposed approaches the distribution network is based on optical fiber links [42,43]. The two approaches to the optical clock distribution are also called [44]:

- microwave frequency transfer
- optical reference generation and distribution.

The first approach is based on the transport, through optical fiber link, of an electrically generated clock signal, while the last method takes advantage of the photonic techniques seen in the previous chapters, for the optical generation and transport of the reference clock. In the following sections, the schematic diagrams and phase stability properties of each technique will be discussed.

6.3.1 Microwave frequency transfer

The simplest method of sharing the frequency reference over optical fiber link is schematically depicted in Figure 6.3. A continuous wave (CW) laser source is amplitude modulated by the clock reference signal through an electro-optical modulator (EOM) and transmitted over the optical link. At the remote node, the light is photo-detected recovering the electrical transmitted clock to be used as local reference signal.

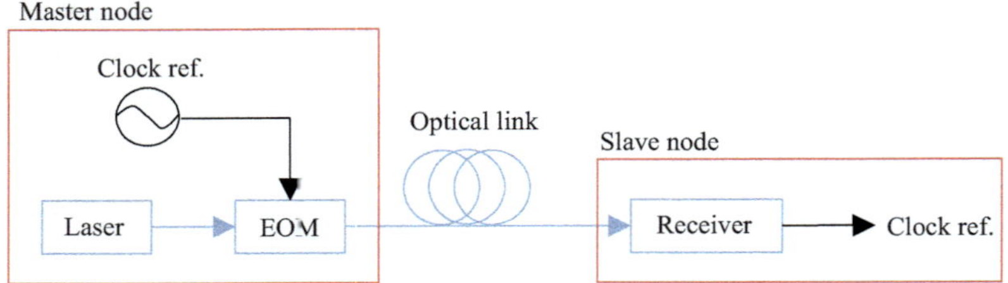

Figure 6.3 Frequency reference is shared over optical fiber link (EOM, electro-optical-modulator)

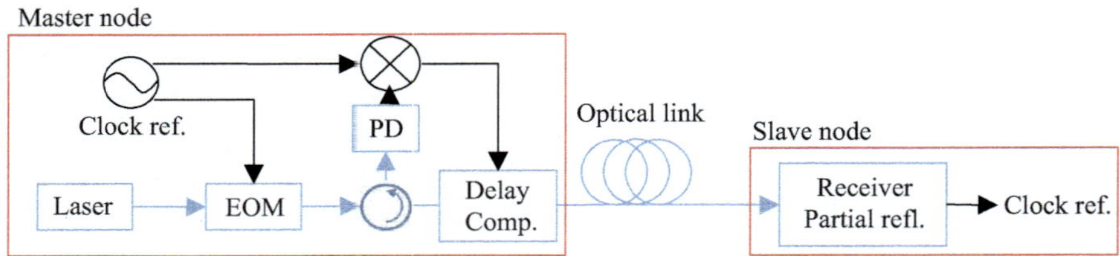

Figure 6.4 Frequency reference is shared over optical fiber link (EOM, electro-optical modulator; PD, photodiode; Delay Comp., delay compensator)

Such a simple scheme has been demonstrated to guarantee an additive fractional phase noise as low as 10^{-14}, with 100 s averaging time [42,43]. The main source of this phase noise is the fiber group velocity variation, induced by temperature variations and vibration, causing respectively long-term and short-term phase instabilities.

In high precision applications [45], as high-resolution radar imaging (but also in metrology or in gravitational waves experiments), such excess phase noise can become a limiting factor. To increase the phase stability of the fiber link, stabilized fiber-optic distribution systems have been developed.

In Figure 6.4, similarly to the scheme proposed in Figure 6.3, an optical carrier is amplitude modulated with the reference signal and sent to the remote site through the fiber link. At the receiver side, a feedback signal is transmitted back to the master node (e.g. through a partially reflective mirror at the receiver), in this way a round-trip signal, affected by the group velocity variation, is available to the master node. The comparison of the generated reference signal with the round-trip delayed replica allows deriving the phase error induced by the optical-fiber link. The derived error signal is then used to control a group delay compensator, such as a thermally controlled spool of fiber or a piezo-electric stretched fiber Bragg grating. Using this technique [46], NASA demonstrated a compensated link with a phase stability better than 10^{-13} with 1 s integration time for a 16 km long fiber link, improving to an impressive 10^{-17} for 10^4 s integration time. Similar approaches [47–49] demonstrated reference stability up to 10^{-15}, with 10^4 s integration time, for optical fiber links up to 86 km.

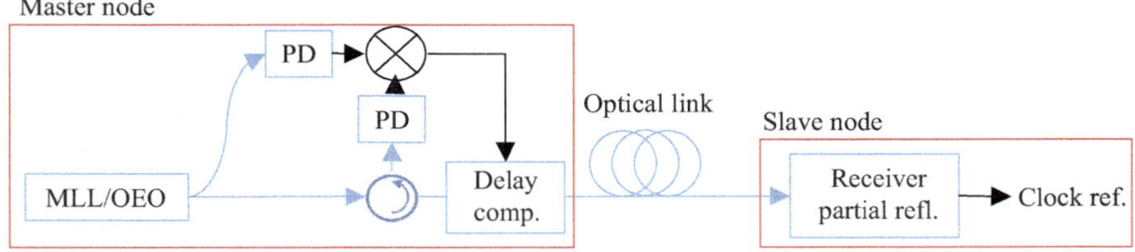

Figure 6.5 Frequency reference is generated and then distributed over optical fiber link (MLL, mode-locked laser; OEO, opto-electronic oscillators; PD, photodiode; Del. Comp., delay compensator)

6.3.2 Optical reference generation and distribution

Although the microwave frequency transfer is an effective and straightforward method for clock distribution, photonics-based microwave frequency generators demonstrate higher phase and frequency stability compared to the electronic counterparts [49] (see Chapter 3).

With this approach, as described in Chapter 3, highly stable mode-locked laser (MLLs) or opto-electronic oscillators (OEOs) are used. The peculiar feature of these sources is that, once photo-detected, they generate a highly phase-stable tone, at a frequency equal to the mode separation of the optical source. As shown in Figure 6.5, this approach can be combined with the frequency stabilization method proposed for the microwave frequency transfer technique, thus further increasing the generated clock phase stability.

Besides allowing an easy distributed synchronization, photonics can also be used to implement a centralized radar network, where the radar signals (and not only the clock signal) are generated in the master node and distributed to the slave nodes through optical fibers. This very recent approach is described in the following section.

6.4 Photonics-based centralized radar network: an experimental approach

As should be clear from the sections above, the centralized coherent radar network is the most performing network architecture since the coherence between the nodes is ensured by the centralized configuration and the fusion process can make use of the raw data from multiple viewpoints, allowing the extrapolation of the maximum information from the observed scenario. On the other hand, this architecture poses critical issues on the distribution of the signals to and from the master node, particularly in relation to the phase stability and to the transmission bandwidth.

As discussed in the previous chapters, photonic technologies can be effectively exploited to generate, transport, and process microwave signals [50,51], guaranteeing a wide system bandwidth, low signal distortion, and immunity to electromagnetic interferences.

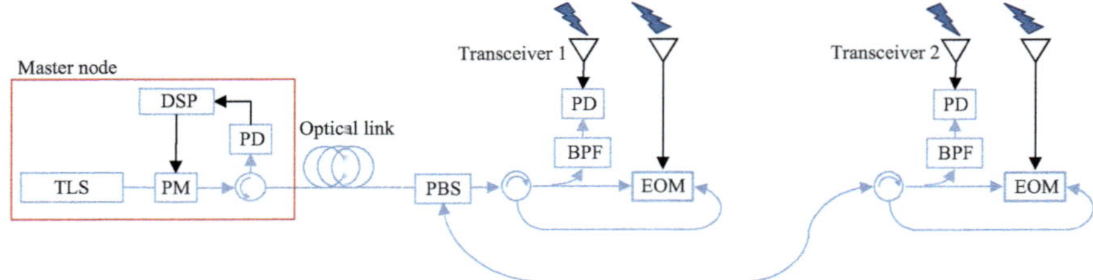

Figure 6.6 Schematic diagram of the fiber-distributed UWB radar based on OTDM (TLS, tunable laser source; PM, phase modulator; PD, photodetector; DSP, digital signal generation and processing; PBS, polarization beam splitter; BPF, bandpass filter; EOM, electro optical modulator)

Exploiting many of the principles seen in those chapters, in [52] an all optical ultra-wideband radar network has been recently demonstrated. To our knowledge, this is the first example of photonics-based centralized coherent radar network. Although it is just a lab prototype, we believe it is worth describing it in more detail.

The system architecture is shown in Figure 6.6. It is a centralized radar architecture, composed of a master node, which contains the signal generation and processing elements, and two remote radar transceiver nodes (the slave nodes). Master and remote nodes are connected by a bi-directional optical fiber link. In the master node, a tunable laser source (TLS) generates a CW signal that is phase modulated with a train of Gaussian pulses generated by the digital signal generation and processing block (DSP) and distributed to the remote radar nodes through the optical fiber link. The pulses are as short as 80 ps, thus reaching the ultra-wide bandwidth of 12.5 GHz. The optical signal is then sent to the remote nodes through a 8-km-long span of single-mode optical fiber. At each remote node, the light from the master node is split into two branches. In the transmitting branch, an optical frequency discriminator is exploited to convert the phase modulation into an amplitude modulation. This device (basically, an optical filter with a steep transmission variation as a function of the signal wavelength) actually transforms the Gaussian phase pulses into ultra-wideband amplitude doublet pulses, exploiting the instantaneous frequency variation given by the phase modulation. In the two remote nodes, the frequency discriminators have opposite signs, so that in one node the Gaussian phase-modulated pulse is transformed into a positive doublet, while in the other it gives a negative doublet, therefore generating two orthogonal waveforms at the two nodes.

After the photodetector, the radar pulses are electrically amplified by the broadband RF front ends and transmitted by the ultra-wideband antennas. In the receiver branch, the light received by the master node is amplitude modulated by means of an EOM, with the echo signal received by the receiving antenna and then sent to the master node for the required signal processing. In the master node, the echo signal received by the two transceivers is discriminated by using a time division multiplexing technique. To this purpose, a proper optical delay, in this case a single-mode fiber spool, is used to separate in time the signal received by the two remote nodes, allowing to share, at the master node side, the same receiver and optical laser source among the two remote transceivers.

Finally, once the time delay between the transmission and reception of the pulses to/from the two remote nodes are known, a simple geometrical method can be used to estimate the position of targets in the XY plane.

This example of photonics-based radar network implements the paradigm of a centralized coherent radar network (see Table 6.1). It is worth underlining that the use of photonics here has allowed the perfect coherence in the signal transmitted by the two nodes, the use of an ultra-broadband radar pulse (with a signal bandwidth >10 GHz), and a centralized approach, leading to an experimental target position detection with a maximum error <2.5 cm in both the range and cross-range directions, despite the easy processing implemented.

This example is limited in the number of orthogonal remote nodes, limited to 2 by the doublet generation technique.

The same group very recently has proposed a novel scheme [53], exploiting wavelength division multiplexing and frequency-disjoint linear frequency modulation to identify orthogonal transmitters. Moreover, photonics-based frequency quadrupling and de-chirping are used to relax the bandwidth requirements on DACs and ADCs. In an experimental validation of the proposed scheme, the researchers have demonstrated a 2×2 MIMO radar network with a bandwidth at each node as large as 4 GHz, and exploiting ADCs with a sampling rate as low as 100 MB/s, reaching a target position resolution of 3.75 cm.

6.5 MIMO processing in photonics-based centralized radar networks

Considering the benefits of photonics in allowing the development of centralized coherent radar networks, in [54] the possibility of applying a specific and powerful MIMO processing in a photonics-based radar network scenario, characterized by widely separated antennas, has been recently taken into account.

The MIMO processing approach steps from the work reported in [13]. As will be detailed in the following, and as already suggested in Chapter 5, the procedure aims at optimizing the SNR of the fused signal by compensating the different propagation delay for each multi-static transmitter–receiver pairs, calculating the corrected echo signal in each point of a finely spaced grid.

As will be seen below, if the node distribution is correctly known and the signals are coherent, this processing is capable of providing a high resolution in both range and cross-range directions. Moreover, under certain hypotheses, even a “super-resolution” can also be achieved (especially in the cross-range direction), exceeding the nominal resolution determined by the radar signal bandwidth.

6.5.1 A simulator for photonics-based MIMO radars

As discussed above, operative centralized coherent MIMO radars still do not exist due to the issues in providing the signal coherence with RF links, and to the strong attenuation in distributing the RF signals on coaxial cables. Therefore, the specifically required fusion algorithms have not been developed yet.

In [54,55] instead, we have considered a photonics-based centralized coherent radar, and we have analyzed what a MIMO processing can bring to the system in terms of detection and localization performance, and which are the critical factors that have to be considered in its design.

A scenario simulator has been first implemented. The simulator considers the exact position in a 2-D plane of the network nodes (distinguishing between transmitting and receiving antennas), and the position of the targets. It can take into account any scenario generically modeled as a configuration of M transmitters and N receivers.

Here we consider the scenario reported in Figure 6.7. The antennas are distributed along a circular arc of 90°, and the target is placed at the center of the circumference, equidistant to every antenna. We use $M=9$ transmitting antennas and $N=8$ receiving antennas, equally spaced along the circular arc. The considered target is an ideal point scatterer at the center of the circular arc, and we assume that all the antennas are able to see it.

The simulator generates the M transmitted signals and calculates the $M \times N$ received signals. The transmitted signals are assumed orthogonal, which can be achieved, for example, by multiplexing them in time, in frequency, or in code.

Let's consider the kth transmitter, emitting the signal s_k. The signal received by the lth receiver can be expressed as:

$$r_{k,l}(t) = A \cdot s_k\big(t - \tau_{k,l}(x,y)\big) \cdot e^{j\left(\theta\left(t-\tau_{k,l}\right)-\theta(t)\right)} + n_{k,l}(t) \tag{6.2}$$

In the expression above, A is an amplitude factor (it depends on the RCS of the scatterer, the transmitted signal, the distance of the target, the antenna gain, the loss

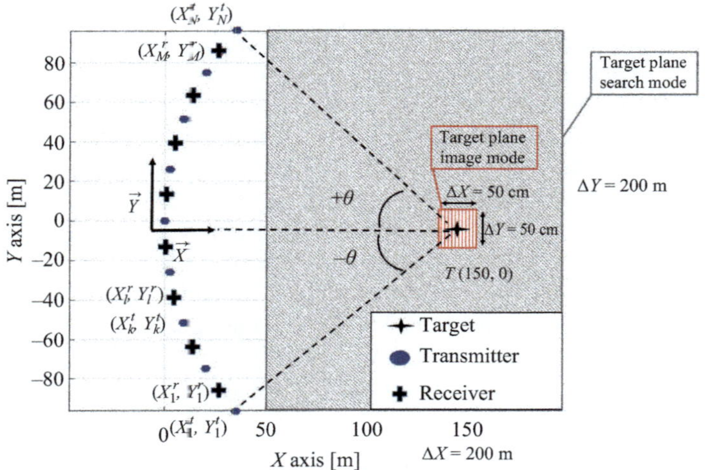

Figure 6.7 Reference scenario

factors, and the carrier wavelength), $\tau_{k,l}$ is the time the signal takes to cover the transmitter–target–receiver path (it depends on the location of the target and of the kth transmitter and lth receiver) for a target placed at the position indicated by the coordinates (x, y). Moreover, $\theta(t)$ is a term taking the phase noise into account, and $n_{k,l}(t)$ is the amplitude noise.

For each possible location (x, y), the simulator calculates the following log-likelihood decision statistic function for considering the MIMO approach [13]:

$$\ln[f(r(t)|(x,y))] = C'\left|\sum_{k=1}^{M}\sum_{l=1}^{N} e^{-j\omega_c\tau_{k,l}}\int r_{k,l}^{b*}(t)s_k^b(t-\tau_{k,l})dt\right| + C'' \quad (6.3)$$

with $\omega_c = 2\pi f_c$ and f_c the carrier frequency, while C' and C'' are constant that do not depend on the scatterer location. The integral term is the correlation between the received and transmitted signal at baseband (indicated by the superscript b). The exponential term compensates for the time delay $\tau_{k,l}$, in practice allowing to sum up coherently the contributions from the different multistatic transmit–receive pairs, thus optimizing the SNR of the function. This term assumes that the signal contributions in the radar network are all coherent, which in our assumptions is guaranteed by the photonic approach.

6.5.2 Simulation results: the potentials of the coherent MIMO processing

The simulations reported in the following have been run considering a carrier frequency of 10 GHz and pulsed radar signals with a bandwidth of 1 GHz. Moreover, in the simulations where the noise has been considered, the amplitude noise has been modeled as an additive white Gaussian noise, while the phase noise has been obtained weighting a white Gaussian noise with the spectral shape of a typical phase noise from optical clocks.

6.5.2.1 Improved resolution capabilities

In a single radar, while the range resolution Δr is defined by the bandwidth B of the transmitted signal ($\Delta r = c/2B$, with c the speed of light), the azimuth (or cross-range) resolution depends on the width of the radiation pattern of the antenna and on the target distance.

In case of a multistatic radar, due to the distributed geometry of the system, first of all it is necessary to define range and cross-range with respect to a common reference system for all the nodes in the network. In the scenario in Figure 6.7, we define the range and cross-range dimensions along the x- and y-axis of the radar network, respectively. Thanks to the geometrical diversity, in the best case (i.e., when the same target is observed from at least two perpendicular viewpoints) the maximum achievable resolution of a noncoherent multistatic radar is equal to $c/2B$ for both range and cross-range.

It is fundamental to point out that, as far as antennas with large beam aperture are used, the radar networks are subject to the detrimental phenomenon of the ghost targets. These are fake targets that appear when more than one target is simultaneously illuminated by all the nodes in the radar network, at all the locations identified by a total distance transmitter–location–receiver equal to those of the real targets.

The discrimination of multiple targets from multiple viewpoints is a typical linear problem. In order to correctly distinguish multiple targets on the scene, it is necessary that the number of observations exceeds the number of targets. In the general case of a MIMO radar network with $M + N$ antennas, the maximum number of correctly distinguishable targets is $M \times N - 1$. For comparison, in a noncoherent radar network with N transceivers, the maximum number of distinguishable targets is $N - 1$, much less than in the MIMO case.

Applying the MIMO processing described by (6.3), it is possible to increase the maximum achievable resolution in the radar network, and to reach the so-called super-resolution [13], that is, a resolution better than the one assured by the signal bandwidth only.

Since the coherent processing of (6.3) requires a significant computational effort, an operative radar network should have two working modes: (1) a standard "search mode" for target detection that does not consider the coherence of the multistatic signals and (2) an "imaging mode" where the coherence is considered and the processing of (6.3) is used to increase the resolution. In the scenario of Figure 6.7, the decision statistic function described above (i.e., the "imaging mode") has been calculated on a squared area of 50 cm per side at the center of the observed scene, and on a grid of positions separated by one-tenth of the carrier wavelength along both range and cross-range directions.

Figure 6.8 shows the range/cross-range maps in "search mode" (Figure 6.8(a), noncoherent processing) and in "imaging mode" (Figure 6.8(b), coherent processing) considering noiseless signals. For calculating the noncoherent processing case, the formula in (6.3) has been used suppressing the exponential term of the delay compensation. Both the algorithms are able to detect the three targets, but in the coherent case the several ambiguities evident in the noncoherent case are suppressed.

Figure 6.9 shows the results considering two targets at a cross-range distance of 3 cm, which is significantly less than the nominal resolution given by the signal bandwidth (in our case, equal to 15 cm). In the absence of noise sources, the two targets are successfully identified applying the coherent MIMO processing. This result demonstrates what is usually called super-resolution [13]. This significant gain in cross-range resolution can be considered as an equivalent of the array factor in sparse array antennas: the larger the number of antenna elements (here, the number of multistatic transmitter–receiver pairs), the narrower the antenna beam (here, the higher the coherence gain).

Figure 6.9 also reports two sets of additional side lobes. These are secondary peaks given by the coherent MIMO processing, and can cause potential ambiguities (false alarms). The number and the height of these side lobes strictly depend on the number of antennas in the radar network: the lower the number of multistatic antennas (i.e., the lower the spatial information), the higher the side lobes.

The spatial information gathered by the radar network is influenced also by the spatial distribution of the antennas (e.g., the aperture in degrees of the circular arc of Figure 6.7), affecting in particular the range resolution. It turns out that

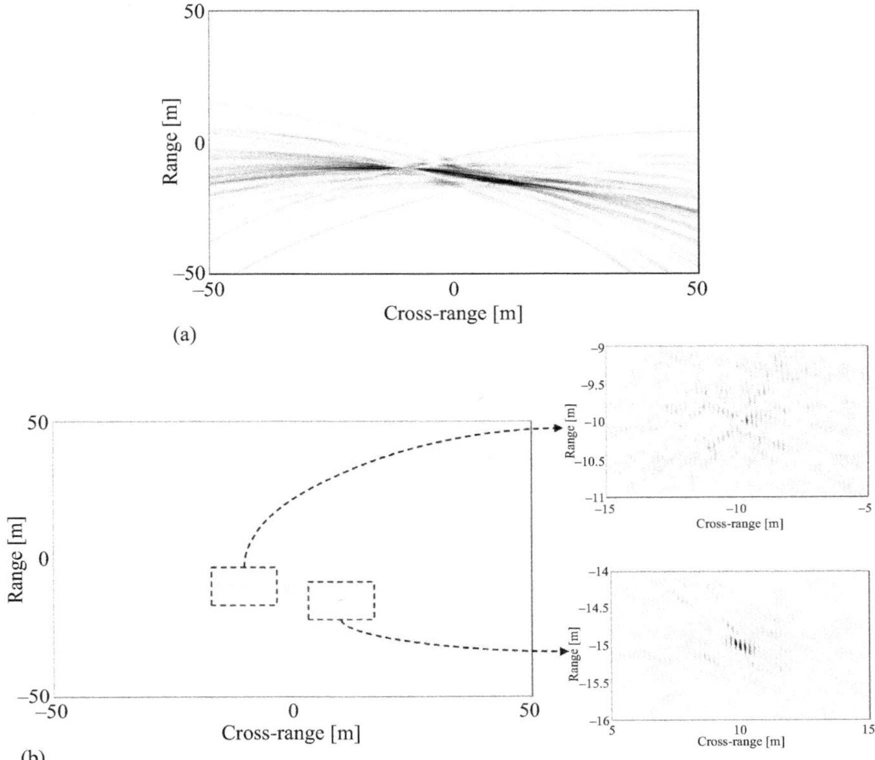

Figure 6.8 Example of range/cross-range map with (a) noncoherent processing and (b) coherent processing

increasing the spatial information given by the different viewpoints reduces the full-width at half-maximum of the target peak lobe [54]. In the example of Figure 6.7, considering the unrealistic distribution of the antennas around a complete circumference, the obtained main lobe is narrowed down to 0.44 cm, demonstrating the super-resolution also in the range profile.

6.5.2.2 Impact of noise

The analysis reported in [54] shows that the amplitude noise has an effect on the extinction ratio of the side lobes with respect to the main lobe, increasing the relative power of the side lobes.

Instead, the effect of phase noise is more complex and has a deeper impact on the detection compared to the amplitude noise.

Considering a centralized coherent radar network, all the nodes share the same coherent photonic RF generator, and the phase noise term in (6.2) is the same for all the $M \times N$ received signals. Here we consider a complex scenario with five closely spaced targets (Figure 6.10), similar to the analysis reported in [54]. First, a low

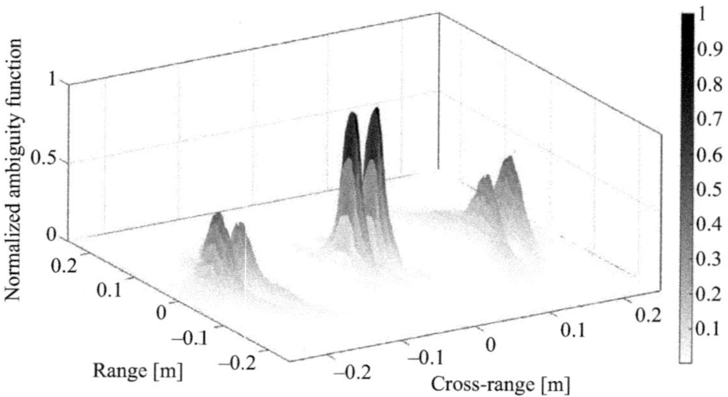

Figure 6.9 Cross-range profile of two targets with distance of 3 cm, detected with coherent processing

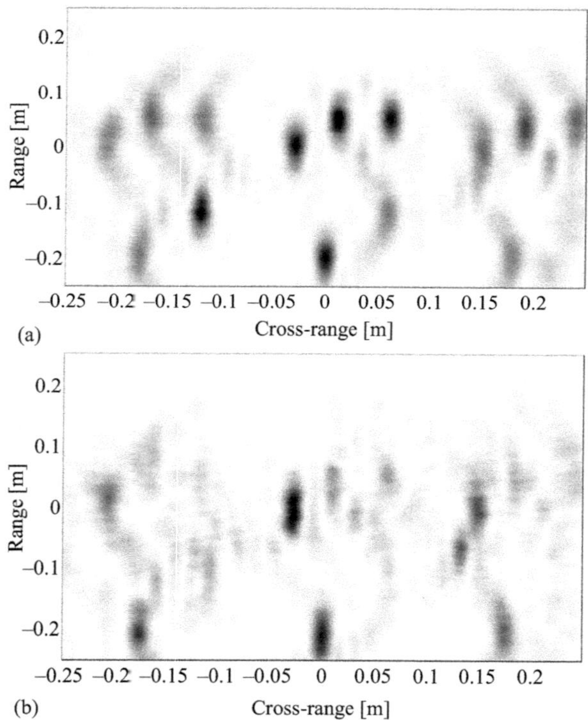

Figure 6.10 Detection of five closely spaced targets distributed using a 9-Tx/8-Rx photonics-based MIMO radar network in presence of phase noise: (a) integrated phase error of 10 mrad and (b) integrated phase error of π rad

phase noise is considered with an integrated phase error of 10 mrad (which is typical of an MLL, a photonics-based clock that is often considered for the realization of photonics-based radars). This phase noise is far smaller than the maximum phase deviation (i.e., 100 mrad) typically required for MIMO processing [56]. In this case, the photonics-based MIMO radar architecture can correctly detect all the five targets in the observed scene (Figure 6.10(a)). Then, a larger phase noise is considered, with an integrated phase error as large as π rad. In this case, which is well above the usually considered limit of 100 mrad, the photonics-based MIMO radar can clearly detect only two out of the five targets present in the scene, with the other three targets becoming comparable with fake side lobes (Figure 6.10(b)). Therefore, the presence of phase noise on the radar signals can cause a reduction of the amplitude of the main lobes, and this, in case of strong phase noise, can even lead to missed detections.

In summary, the simulative analysis of a coherent centralized photonics-based MIMO radar confirms the huge potential of MIMO detection, which is capable of improving the detection of complex scenarios with a high resolution in range and cross-range, with a good robustness against amplitude and phase noise, and the additional capability for reaching a super-resolution.

6.6 Conclusions

The discussions reported in this chapter demonstrate that photonics is going to be an enabling technology for the development of coherent radar networks:

- Exploiting optical fibers, it is possible to synchronize all the nodes in a network, even when the nodes are widely separated.
- Optical fibers also allow to implement a centralized network, thanks to the enormous transmission bandwidth.
- The centralized network can be based on highly stable photonics-based clocks and signal generators for further improving the network performance.
- The precision in the synchronization permits the development of MIMO processing, leading to a significant improvement in the understanding of the observed scenario (range and cross-range resolution, super-resolution, detection of stealth objects, detection of slow moving targets, etc.).

The recent experimental activities reported in the chapter set a strong technical basis for further developments and clearly suggest that the potential of photonics in radar networks is going to be explored—and exploited—more and more in the next future.

Moreover, the deployment of optical fibers (e.g., as local data networks, or as metropolitan telecommunication networks) is becoming increasingly common in our society, and this can further foster the birth of operative photonics-based radar networks.

All the advances above describe a bright future for coherent radar networks, enlightened by the use of photonics.

References

[1] Willis, N. J. *Bistatic Radar*. 2nd. edn. Raleigh: SciTech Publishing Inc., 2005.

[2] Willis, N. J. and Griffiths, H. *Advances in Bistatic Radar*. Edison: SciTech Publishing Inc., 2007.

[3] Hanle, E. Survey of bistatic and multistatic radar. *IEE Proceedings on Communications, Radar and Signal Processing*, Vol. 133, 7, pp. 587–595, 1986.

[4] Chernyak, V. S. *Fundamentals of Multisite Radar Systems: Multistatic Radars and Multistatic Radar Systems*. 1st edn. London, UK: CRC Press, 1998.

[5] Baker, C. J. *An Introduction to Multistatic Radar*. NATO-SET 136 Lecture Series Multistatic Surveillance and Reconnaissance: Sensor, Signal and Data Fusion, 2009.

[6] Baker, C. J. *Multistatic Radar Processing and Systems*. NATO SET-136 Lecture Series on Multistatic Surveillance and Reconnaissance: Sensor, Signals and Data Fusion, 2009.

[7] Baker, C. J. and Hume, A. L. Netted radar sensing.. *Proceedings of the CIE International Radar Conference*, Beijing, China, pp. 110–114, 2001.

[8] Robey, F. C., Coutts, S., Weikle, D., *et al. MIMO* radar theory and experimental results. *Proceedings of the 37th ASILOMAR 2004 Conference on Signals, Systems and Computers*, Pacific Grove, CA, USA, pp. 300–304, 2004.

[9] Bekkerman, I. and Tabrikian, J. Target detection and localization using MIMO radars and sonars. *IEEE Transactions on Signal Processing*, Vol. 54, pp. 3873–3883, 2006.

[10] Xu, L., Li, J., and Stoica, P. Adaptive techniques for MIMO radar. *Proceedings of the 14th IEEE Workshop on Sensor Array and Multi-Channel Processing*, Waltham, MA, USA, July 2006.

[11] Fishler, E., Haimovich, A., Blum, R. S., Cimini, L. J., Chizhik, D., and Valenzuela, R. A. Spatial diversity in radars – models and detection performance. *IEEE Transactions on Signal Processing*, Vol. 54, pp. 823–838, 2006.

[12] Lehmann, N., Haimovich, A., and Valenzuela, R., *MIMO* – Radar application to moving target detection in homogeneous clutter. *Proceedings of the 14th IEEE Workshop on Sensor Array and Multi-Channel Processing*, Waltham, MA, USA, July 2006.

[13] Lehmann, N. H., Haimovich, A. M., Blum, R. S., and Cimini, L., High resolution capabilities of MIMO radar. *Proceedings of the 40th ASILOMAR 2006 Conference on Signals, Systems and Computers*, Pacific Grove, CA, USA, November 2006.

[14] Godrich, H., Haimovich, A. M., and Blum, R. S. Cramer-Rao bound on target localization estimation in MIMO radar systems. *Proceedings of the 42nd Annual Conference on Information Sciences and Systems (CISS)*, Princeton, NJ, USA, March 2008.

[15] Skolnik, M. I. *Introduction to Radar Systems*. 3rd. edn. New York: McGraw-Hill Book Company, 2008.

[16] Weiss, M. Synchronization of bistatic radar systems. *IEEE 2004 International Geoscience and Remote Sensing Symposium*, Anchorage, Alaska, USA, pp. 1750–1753, 2004.

[17] Yang, Y. and Blum, R. S. Phase synchronization for coherent MIMO radar: algorithms and their analysis. *IEEE Transactions on Signal Processing*, Vol. 59, 11, pp. 5538–5557, 2011.

[18] Hurley, S. M., Tummala, M., Walker, T. O., and Pace, P. E., Impact of synchronization on signal-to-noise ratio in a distributed radar system. *2006 IEEE/SMC International Conference on System of Systems Engineering,* Los Angeles, CA, USA, April 2006. 10.1109/SYSOSE.2006.1652311.

[19] Sundararaman, B., Buy, U., and Kshemkalyani, A. D. Clock synchronization for wireless sensor networks: a survey. *Ad Hoc Networks*, Vol. 3, 3, pp. 281–323, 2005. https://doi.org/10.1016/j.adhoc.2005.01.002.

[20] Li, Q. and Rus, D. Global clock synchronization in sensor networks. *IEEE Transactions on Computers*, Vol. 55, pp. 2, 214–226, 2006,.

[21] Wang, W. Q. and Shao, H. Z. Performance prediction of a synchronization link for distributed aerospace wireless systems. *The Scientific World Journal*, Hindawi, 2013. doi:10.1155/2013/159742.

[22] Goh, A. S., Preiss, M., Stacy, N. J. S., and Gray, D. A., Bistatic SAR experiment with the Ingara imaging radar. *IET Radar, Sonar and Navigation*, Vol. 4, 3, pp. 426–437, 2010.

[23] Weiss, M. *Aspects of Sensor Networks*. 2012. RTO-EN-SET-157.

[24] Soame, T. A. and Gould, D. M. Description of an experimental bistatic radar system. *Proceedings of IEE International Radar Conference*, Vol. 281, pp. 12–16, 1987.

[25] Dunsmore, M. R. B. Bistatic radars for air defense. *Proceedings of IEE International Radar Conference*, London, UK, Vol. 281, 1987.

[26] Wendler, M. Results of a bistatic airborne SAR experiment. *Proceedings of the International Radar Conference*, Adelaide, Australia, pp. 247–253, September 2003.

[27] Espeter, T., Walterscheid, I., Klare, J., Brenner, A. R., and Ender, J. H. G., Bistatic forward-looking SAR: results of a spaceborne airborne experiment. *IEEE Geoscience and Remote Sensing Letters*, Vol. 8, 4, pp. 765–768, 2011.

[28] Walterscheid, I., Espeter, T., Brenner, A. R., *et al.* Bistatic SAR experiments with PAMIR and TerraSAR-X-setup, processing, and image results. *IEEE Transactions on Geoscience and Remote Sensing*, Vol. 48, 8, pp. 3268–3279, 2010.

[29] Auterman, J. L. Phase stability requirements for a bistatic SAR. *Proceedings of the IEEE National Radar Conference*, Atlanta, USA, pp. 48–52, 1984.

[30] Weiss, M. Time and phase synchronization aspects for bistatic SAR systems. *Proceedings of the European Synthetic Aperture Radar Conference*, Ulm, Germany, pp. 395–398, 2004.

[31] Wang, W. Q. *Multi-Antenna Synthetic Aperture Radar*. London, UK: CRC Press, 2013.

[32] Choi, B. J., Liang, H., Shen, X., and Zhuang, W., DCS: distributed asynchronous clock synchronization in delay tolerant networks. *IEEE Transactions on Parallel and Distributed Systems*, Vol. 23, 3, pp. 491–504, 2012.
[33] Leng, M. and Wu, Y. C. Distributed clock synchronization for wireless sensor networks using belief propagation. *IEEE Transactions on Signal Processing*, Vol. 59, 11, pp. 5404–5414, 2011.
[34] Krieger, G., and Younis, M., Impact of oscillator noise in bistatic and multistatic SAR. *Proceedings of the IEEE International Geoscience and Remote Sensing Symposium*, Seoul, South Korea, pp. 1043–1046, July 2005.
[35] Ubolkosold, P., Knedlik, S., and Loffeld, O. Estimation of oscillator's phase offset, frequency offset and rate of change for bistatic interferometric SAR. *Proceedings of the European Synthetic Aperture Radar Conference*, Dresden, Germany, pp. 1–4, 2006.
[36] Zhang, X., Li, H., and Wang, J. The analysis of time synchronization error in bistatic SAR system. *Proceedings of the IEEE International Geoscience and Remote Sensing Symposium*, Seoul, South Korea, pp. 4619–4622, July 2005.
[37] Wang, W.Q., Ding, C. B., and Liang, X. D. Time and phase synchronisation via direct-path signal for bistatic synthetic aperture radar systems. *IET Radar, Sonar and Navigation*, Vol. 2, 1, pp. 1–11, 2006.
[38] Eineder, M. Ocillator clock drift compensation in bistatic interferometric SAR. *Proceedings of the IEEE International Geoscience and Remote Sensing Symposium*, Toulouse, France, pp. 1449–1451, July 2003.
[39] Younis, M., Metzig, R., and Krieger, G. Performance prediction of a phase synchronization link for bistatic SAR. *IEEE Geoscience and Remote Sensing Letters*, Vol. 3, 3, pp. 429–433, 2006.
[40] Retzer, G. A concept for signal processing in bistatic radar. *IEEE International Radar Conference*, Arlington, VA, USA, pp. 288–293, 1980.
[41] Kanno, A. and Yamamoto, N. Radio over fiber network technology for millimeter-wave distributed radar systems. *Proceedings of SPIE 10559, Broadband Access Communication Technologies XII,* San Francisco, CA, USA, 29 January 2018. 10.1117/12.2287731.
[42] Masaki, A. Time and frequency transfer and dissemination methods using optical fiber network. *Proceedings of the 2005 IEEE International Frequency Control Symposium and Exposition*, Vancouver, BC, Canada.
[43] Predehl, K., Grosche, G., Raupach, S. M. F., *et al.* A 920-kilometer optical fiber link for frequency metrology at the 19th decimal place. *Science*, Vol. 27, pp. 441–444, 2012.
[44] Foreman, S. M., Holman, K. W., Hudson, D. D., Jones, D. J., and Yed, J., Remote transfer of ultrastable frequency references via fiber networks. *Review of Scientific Instruments*, Vol. 78, 2, 021101, 2007.
[45] Krisher, T. P., Maleki, L., Lutes, G. F., *et al.* Test of the isotropy of the one-way speed of light using hydrogen-maser frequency standards. *Physical Review*, Vol. D 42, 2, 1990.
[46] Shillue, B. *ALMA LO Distribution Round Trip Phase Correction*. 2002. ALMA Memo# 443, 731.

[47] Narbonneau, F., Lours, M., Bize, S., Clairon, A., and Santarelli, G., High resolution frequency standard dissemination via optical fiber metropolitan network.. *Review of Scientific Instruments*, Vol. 77, 6, 064701, 2006.

[48] Daussy, C., Lopez, O., Amy-Klein, A., *et al.* Long-distance frequency dissemination with a resolution of 10–17. *Physical Review Letters*, Vol. 94, 20, 2005.

[49] Celano, T. P., Stein, S. R., Gifford, G. A., Mesander, B. A., and Ramsey, B. J., Sub-picosecond active timing control over fiber optic cable. *Proceedings of the 2002 IEEE International Frequency Control Symposium and PDA Exhibition Institute of Electrical and Electronics Engineers*, Piscataway, NJ, USA, pp. 510–516, 2002. 10.1109/FREQ.2002.1075937.

[50] Yao, J. Microwave photonics. *IEEE Journal of Lightwave Technology*, Vol. 27, 22, pp. 314–335, 2009.

[51] Capmany, J. and Novak, D. Microwave photonics combines two worlds. *Nature Photonics*, Vol. 1, 6, pp. 319–330, 2007.

[52] Fu, J. and Pan, S. A fiber-distributed bistatic ultra-wideband radar based on optical time division multiplexing. *IEEE 2015 International Topical Meeting on Microwave Photonics*, Paphos, Cyprus, 2015.

[53] Zhang, F., Gao, B., and Pan, S Photonics-based MIMO radar with high-resolution and fast detection capability. *Optics Express*, Vol. 26, 13, pp. 17529–17540, 2018.

[54] Lembo, L., Ghelfi, P., and Bogoni, A. Analysis of a coherent distribute MIMO photonics-based radar network. *Proceedings of the European Radar Conference (EuRAD)*, Madrid, Spain (accepted).

[55] Bogoni, A., Lembo, L., Serafino, G., Ghelfi, P., and Scotti, F., Microwave photonics in radar. *Proceedings of the IEEE Photonics Conference (IPC)*, Reston, VA, USA (accepted).

[56] Pasya, I. and Kobayashi, T. Detection performance of M-sequence-based MIMO radar systems considering phase jitter effects. *Proceedings of the 2013 IEEE International Symposium on Phased Array Systems and Technology*, Waltham, MA, USA, 2013. 10.1109/ARRAY.2013.6731861.

Chapter 7
Photonics in electronic warfare systems

Daniel Onori[1] *and Paolo Ghelfi*[2]

7.1 Chapter organization and key points

In Chapter 2, we have seen the requirements for electronic warfare (EW) systems, the current solutions based on standard electronic technologies, and their limits.

Here, we analyze what photonics can implement in this field. In fact, this chapter takes into account the several photonics-based solutions proposed so far for the electronic warfare. In particular, these solutions focus on the following applications: microwave photonic links (MPLs), instantaneous frequency measurement (IFM) systems, channelized receivers, and scanning receivers. In the case of the scanning receiver, we also report on a field trial run in a naval scenario.

It is worth underlining here that most of the photonics-based solutions proposed so far are still at the level of research results. In fact, in order for photonics to be a serious contender in radio-frequency (RF) systems, it needs to demonstrate either a quantum leap in performance or completely new capabilities in comparison with digital electronics. As can be read in the examples below, several photonics-based solutions are coming close to this point. The reader can therefore find in the following the seeds of what we expect to see soon in the real application fields.

7.2 Photonics potentials in EW systems

In the last decade, the scientists involved in the field of electronic warfare have started exploring the use of microwave photonics in defense systems to solve specific problems that conventional electronics could not tackle.

Following the description of the different tasks of electronic warfare, as reported in Section 2.2, photonics has been proposed for implementing several functions as listed below [1]. In the rest of this chapter, we will then focus on those that have gathered the highest interest.

[1]Energie Matériaux et Télécommunications, Institut National de la Recherche Scientifique (INRS), Québec, Canada

[2]National Laboratory of Photonic Networks and Technology (PNTLab), Consorzio Nazionale Interuniversitario per le Telecomunicazioni (CNIT), Italy

7.2.1 Electronic protection

In order to avoid the saturation of the EW receivers due to hostile jammers or unintentional interferences (e.g., strong civilian wireless signals), the use of flexible photonics-based RF filters has been proposed [2]. These filters can be easily tuned to filter out the powerful low-priority signals, so that high-priority (i.e., hazardous) signals that are typically weak (with low probability of intercept) can be distinguished by the acquisition system.

7.2.2 Electronic support

Under this category, we can list all the photonics-based solutions for improving the detection of RF signals in a dense electromagnetic environment. These include the implementations of wideband receivers realizing a photonics-based tunable down-conversion (either as a channelized receiver or as a scanning receiver, see Sections 7.5 and 7.6), or using a photonics-assisted analog-to-digital converter (ADC) to digitize broader spectrum portions (see also Section 3.4); direction finders or frequency measure systems exploiting photonics; MPLs (also called radio-over-fiber (RoF) systems, see Section 3.5) with extended dynamic range to allow the remotization of antennas; broadband beamforming systems based on true-time delays enabled by photonics (see also Section 3.7).

7.2.3 Electronic attack

In order to deceive and defeat the hostile radars, schemes have been presented using photonics for realizing an effective jamming [3]. This is realized by detecting the hostile RF signals, and retransmitting them after having shifted their frequency by means of broadband true-time delays. This way, the hostile radars will misinterpret the velocity of the target, losing its tracking. Another feature that photonics can bring to jammers is the capacity of simultaneously transmitting and receiving [4]. In fact, jammers need a very high isolation between the transmitter and the receiver in order to avoid the signal ringing. Standard jammers do not have such an isolation and need to resort to chopping. Unfortunately, chopping can be easily recognized by the hostile radar. Photonics has been reported allowing the full duplex operation, increasing the safety of the Electronic Attach (EA) system.

In the following, we report on the main photonics-based solutions that have been proposed for EW applications. As can be seen, the most targeted task is the electronic support for which a field trial is also reported.

7.3 Microwave photonic links

The motivations and basic principles of MPLs have been introduced in Section 3.5 from a general point of view. Here, we focus in more detail on the use of MPLs in the EW field.

In several defense applications, the possibility of remoting an antenna from its base station is particularly desirable because it would allow the flexibility of placing the two subsystems where is more convenient in terms of encumbrance or power consumption. Moreover, separating the base station from the antenna would also allow to centralize the processing of the information gathered by several antennas distributed around the platform (e.g., an airplane), helping in classifying the possible threats [5]. Since the attenuation of a good coaxial cable at 18 GHz is usually >1 dB/m, it is clear that in standard RF systems the distance between the antenna and its base station must be within few meters, or the link budget of the system becomes unbearable.

On the other hand, the advances in fiber optics driven by the communications sector has rapidly provided the defense systems with viable and affordable solutions for the transportation of broadband RF signals on fiber: this systems are usually referred to as MPLs or RoF systems. In fact, fiber has a typical attenuation as low as 0.2 dB/km, making the propagation losses negligible for distance within few tens of kilometers. Moreover, the fiber transmission is free from electromagnetic interferences. Finally, fibers (even the most ruggedized) are largely more flexible, narrower, and lighter than coaxial cables, giving the RoF systems an enormous flexibility concerning installation.

7.3.1 MPLs based on intensity modulation and direct detection

Figure 7.1 shows the scheme of the most common RoF system. It is based on the paradigm of intensity modulation and direct detection (IM-DD). Considering the transfer of the signal from the base station to the antenna, the broadband RF signal is translated in the optical domain by modulating an optical carrier in a Mach–Zehnder modulator (MZM) biased at its quadrature. Then, the optical signal is transported through optical fiber to the RF base station, where a photodiode (PD) moves the signal back to the RF domain. Thanks to the components made available by the optical communication market, RoF systems covering the RF frequency range from <2 GHz up to >40 GHz are commercially available. Table 7.1 reports the typical characteristics of the main system components and the related RoF performance.

The performance limits of RoF systems are primarily given by noise and distortion. Noise limits the minimum microwave signal level that can be detected, while distortion (i.e., lack of linearity) limits the maximum microwave signal power that can be transmitted. The spur-free dynamic range (SFDR), that is, the

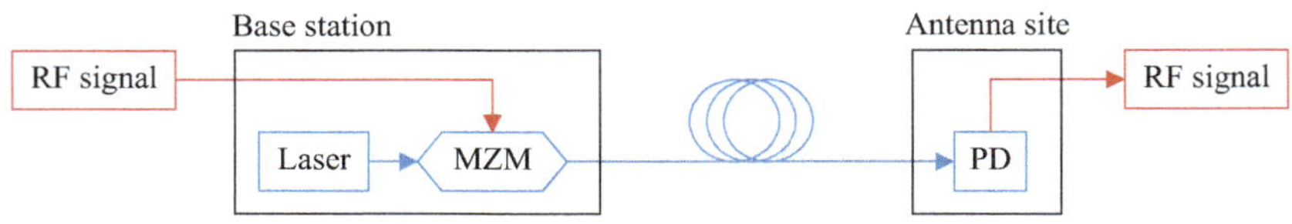

Figure 7.1 Scheme of a common RoF system for remoting antenna and RF base station

Table 7.1 Typical parameters of RoF components and performance of the RoF link [5]

Component parameters		Link RF performance	
Laser power	20 dBm	Link gain	–16 dB
Laser relative intensity noise	–160 dBc/Hz	Noise figure	32 dB
Modulator $V\pi$	5 V	Third-order intercept point	23 dBm
Modulator insertion loss	4 dB	SFDR	110 dB·Hz$^{2/3}$
Photodiode responsivity	0.8 A/W	P1dB	14 dBm
Link optical loss	2 dB	Compressive dynamic range	155 dB/Hz

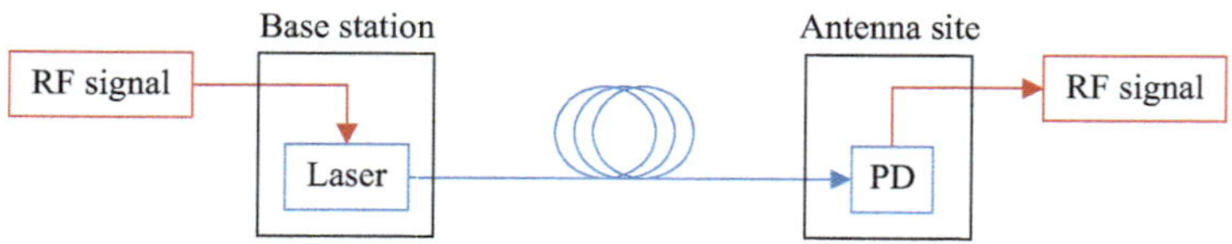

Figure 7.2 Scheme of a microwave photonic link based on the direct modulation of the laser

difference between minimum and maximum signals, is therefore one of the key figures of merit [6].

Although the performance reported in Table 7.1 is sufficient for several EW applications (as the electronic support measures and the electronic intelligence), in some more demanding cases (as for the radar warning receivers) the typical value for SFDR can be not enough.

To extend the use of MPLs to the most demanding applications, few technical solutions have been investigated, as will be detailed in the following. These techniques aim at either optimizing the performance of the IM-DD approach (by predistorting the RF signal to compensate the nonlinear response of the link or by using a differential scheme to cancel out the nonlinearity), or at implementing a substantially different scheme exploiting phase modulation and coherent detection.

Before describing these high-performance techniques, it is worth mentioning here about the MPLs based on the direct modulation of the laser, as sketched in Figure 7.2. In this approach, the RF signal is loaded on the laser intensity by directly modulating the driving current of the laser itself, exploiting the linear current/intensity characteristic of the laser diodes. Then, the intensity modulation is easily detected by a PD, recovering the original RF signal. As it is clear, this scheme has the advantage of saving costs since it avoids using an external modulator. Moreover, it also saves on the RF power needed to transfer the modulation on the optical carrier. In fact, while the MZM has a typical $V\pi$ (i.e., the voltage variation at the RF port of the MZM required to pass from the minimum to the maximum transfer) in the order of few volts peak-to-peak, the laser current can be driven with only few mA.

Unfortunately, the direct modulation MPLs also show significant shortcomings. First, the available modulation bandwidth (BW) of the laser driving current is usually limited to few GHz. Some specific lasers have been developed specifically for fast direct modulation applications in the digital communication field, so they show a BW approaching 20 GHz, but they also have a limited modulation extinction ratio (ER) that is sufficient only for on–off keying signal distribution. Then, the direct modulation of the driving current of a laser induces a modulation of the refractive index in its active region, which translates into a frequency chirping of the directly modulated laser. If the launched optical signal travels into long fiber spans, the chirp is translated into fluctuations of propagation time induced by the fiber chromatic dispersion. At the end, the received RF signal is subject to an additional timing jitter that can become detrimental. Therefore, the direct modulation scheme is less performing than the external modulation ones, and its use is limited to low cost and low performance applications.

7.3.2 MPL linearization by predistortion

One of the main causes for the limited SFDR in the MPLs based on IM-DD is the nonlinear response of the MZM. In fact, this component transfers the time-dependent behavior of the RF signal into a time-dependent modulation of the laser optical power, following a sinusoidal characteristic centered at the quadrature point (the point of maximum linearity). If the input RF signal is small, the transfer is very linear; but if the RF signal amplitude gets larger, the sinusoidal characteristic comes into play, and higher harmonics (in particular, the third harmonic) become visible.

The nonlinear characteristic of the MZM can be compensated for by adding a predistorter at its input RF port (as sketched in Figure 7.3), so that the final transfer function on the MPL has an increased linearity. A reduction of up to 20 dB of the third-order distortion is usually achievable [6].

A relevant example of predistortion is reported in [7]: it works in the band 6–12 GHz, allowing an increase of 6 dB in the SFDR and of about 9 dB in the output power. The analysis in [7] also considers the effects of all the MPL components on the linearization given by the predistorter. It turns out that the benefit of the linearization is dependent on several parameters, as the laser relative intensity noise (RIN), the $V\pi$ of the MZM, the presence of an EDFA, and the gain of the

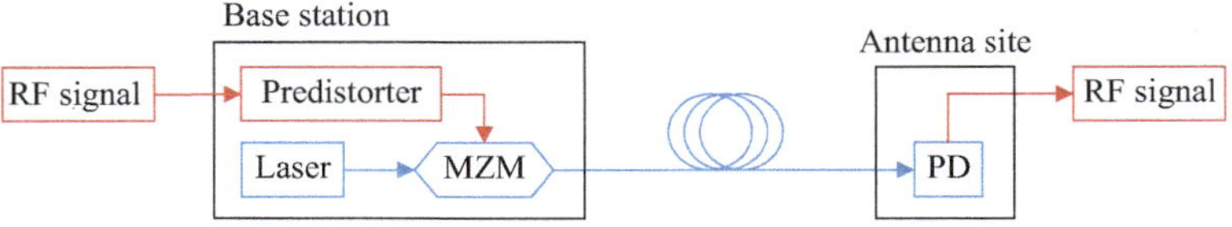

Figure 7.3 Scheme of a microwave photonic link with a predistorter for increasing the linearity

predistorter. Interestingly, the presence of a linearizing predistorter can be even detrimental if the link is characterized by a strong optical power at the PD. In fact, under this condition, the unlinearized system is limited by the shot noise only, but the linearized system could add a significant amplified thermal noise at the input.

7.3.3 Differential transmission and detection

Another approach to increase the dynamic range is to "clear" the signal from the noise at low signal levels. In these signal conditions, the main limitation comes from the RIN of the laser.

In order to suppress the RIN, it is possible to exploit the differential modulation and detection. In fact, any MZM potentially has two output ports since its structure is composed of a 1×2 splitter followed by a 2×2 coupler. The two ports are reciprocal, that is, the optical signals at their outputs bring the RF modulation signal with opposite sign. In other words, the modulation sidebands at the output of the two ports are 180° out of phase. Most of the MZMs are realized connecting one output port only, but a dual output MZM can be easily produced as well. With a dual port MZM, the RF signal can be loaded simultaneously on the two output optical signals, and transmitted along two parallel fibers to a receiver. The receiver can be conveniently composed of two PDs in balanced configuration, so that the current at their outputs are subtracted. Therefore, the common noise (in particular, the RIN) is completely cancelled out [8], but the RF signal turns out to sum up. As it is clear, the balanced detection also allows increasing the amplitude of the recovered RF signal, further contributing at increasing the SNR [9].

Clearly, the advantages of this solution come at the cost of having two fibers with matched length for transmitting the signal, and two PDs (instead of one) in balanced configuration.

7.3.4 Other methods for extending the linearity of IM-DD PMLs

Several other approaches have been presented for extending the dynamic range of an MPL, resorting to complex setups or costly components. One possible way is to increase the link gain by using high power lasers [10] or optical amplification [11]. Other methods aim at improving the signal-to-average power ratio by suppressing the optical carrier or biasing the MZM at a low transmission point [12], showing interesting results when optical amplification is used. An excellent summary of performance limits of the conventional MPLs can be found in [10].

However, for some applications the dynamic range achieved by these approaches is still not large enough. A method to significantly improve further the dynamic range has been proposed in [13], based on a dual MPL architecture and a fast switch. Two optical links are implemented with different sensitivities, and the fast switch sends the signal to the most correct link, driven by a fast circuit that senses the input RF power. This complex and fascinating approach reaches a nominal "synthetic" dynamic range of 100 dB. Actually, it "only" works alternatively on either of two ranges, each with a dynamic range of 50 dB.

7.3.5 MPLs based on phase modulation and coherent detection

In order to transfer an RF signal onto an optical carrier, it could be convenient to use a phase modulation [14,15]. In fact, differently from the MZMs that have an intrinsically nonlinear transfer function, the phase modulator is linear. This should ensure a very large SFDR. The problem comes when the optical signal must be detected to recover the original RF signal. In fact, it is not possible to recover it by simply detecting the optical signal in a PD since this will return only the envelope of the optical power, which is constant in the case of phase modulation. Instead, a phase demodulator is needed, which can be realized by means of a coherent receiver.

Figure 7.4 reports the scheme of an MPL based on phase modulation and coherent detection. The scheme takes into account a second laser at the same wavelength of the optical carrier, working as optical local oscillator (LO). The LO is coupled to the optical signal into a 90° hybrid coupler (HC), which provides at its output the four combinations of signal and LO with phase shifts of 0°, 90°, 180°, and 270°. The outputs are connected 2 by 2 to two balanced photodiodes (BPDs), which finally return the in-phase (I) and quadrature (Q) components of the original RF signal. Then, the I/Q signals are sampled by two ADCs, and successively the RF signal is digitally recovered by calculating the numerical signal $S = I + jQ$.

As it is clear from this description, the output signal of this MPL is not the original RF signal, but its digitized version. Therefore, the MPL approach based on phase modulation and coherent detection can be applied only if the analog RF signal is not required again at the receiver.

In this MPL approach, the LO must be perfectly coherent with the optical carrier. This can be achieved by using exactly the same laser, but this solution is difficult to implement if the input and output of the MPL are broadly separated (as is usually the case). The most exploited method is to lock the LO to the optical carrier either through injection locking or through an optical PLL [16,17].

It is important to underline that the use of an optical LO allows a strong improvement of the MPL efficiency since it can provide high power at the BPDs, and hence recover strong I/Q signals without transmission and component losses.

As expected, the reported results show a strong reduction of third-order distortion with respect to IM-DD systems, and a significant improvement of the SFDR, which can surpass 150 dB·Hz$^{2/3}$ [17].

As said, the description above considers a LO at the same wavelength as the optical carrier, thus realizing a down-conversion to baseband. The maximum signal BW is therefore limited by the accepted BW of the ADCs. The scheme can be

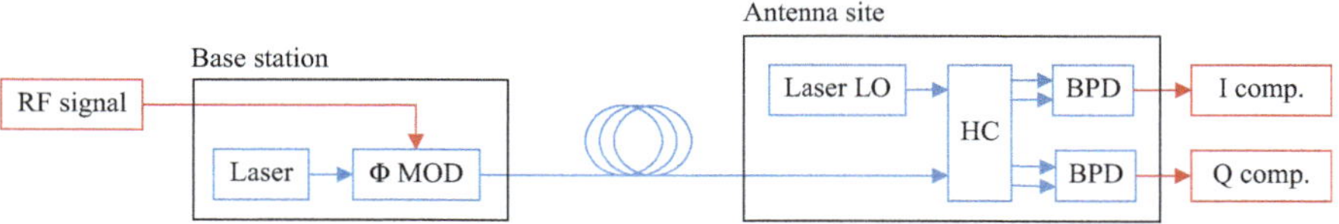

Figure 7.4 Scheme of a microwave photonic link based on the phase modulation of the laser and coherent detection

easily extended to implement a coherent down-converter, provided that the LO has the correct detuning from the optical carrier. In [15], this has been realized by creating the shifted LO from the optical carrier by modulation, and the coherent optical link has been tested up to 40 GHz.

7.3.6 MPLs for direction finding

As a collateral application of antenna remoting, MPLs have been proposed also for simplifying the implementation of direction finding systems [5]. In fact, direction finding is usually implemented by long baselines and remoted antennas, detecting the direction of arrival of a signal analyzing the phase shift registered between the antennas.

As described above, it is evident that exploiting MPLs for this application induces a reduction of size, weight, and power consumption with respect to standard RF transmission through coaxial cables.

7.4 Instantaneous frequency measure systems

In EW applications, it can be practical to exploit specialized receivers to run specific analyses of the radar signal environment, exploiting the strengths of each type of receiver to optimize the situation awareness and reduce the processing load of the full set of receivers. For example, the measurement of the unknown received signals frequency is one task that can be performed more efficiently by a specialized IFM receiver, before letting further processing to another specialized receiver.

The typical requirements for an IFM receiver include a wide BW of operation (at least 2–18 GHz), high resolution, and near real-time response. A microwave receiver, however, usually operates in a very narrow frequency band. So, photonics-assisted IFM receivers have been proposed as promising solutions, thanks to their broad BW, near real-time measurement, large measurement range, low loss, and small size [18–21].

The most straightforward technique for implementing an IFM receiver is the channelized approach. In [18], the received RF signal is modulated on a laser, the modulated spectrum is then split and applied to bank of filters (e.g., a grating array), and an array of PDs detects the power associated to each filtering window of the filter bank. The system demonstrated in the cited work is capable of resolving signals in the 2–18 GHz range into 2 GHz bands in a simple optical topology, whilst maintaining a high probability of intercept. It is worth underlining that with this kind of receiver the specifics of the RF signal such as precise frequency and phase information is lost. These details, however, are intended to be obtained from other signal processing modules.

In [19], the channelized approach is improved by considering a laser comb instead of the single light source, and adding an etalon with a free spectral range slightly different from the comb spacing, so that a sort of interferometric analysis can be implemented, strongly increasing the precision of the frequency estimation.

More recently, few proposals have appeared to be particularly suitable for IFM system, thanks to the reported performance.

In [20], the used technique measures the frequency of the microwave signal by means of a frequency-to-power mapping, so that the microwave frequency is simply estimated by measuring an optical power. In more detail, the cited work proposes a simple microwave photonic system consisting of a two-tap photonic microwave filter pair with complementary frequency responses. One of the filters is a low-pass filter (LPF), and the other is a bandpass filter, both obtained by using a polarization modulator and two sections of polarization maintaining fiber. Thanks to the complementary nature of the frequency responses of the photonic microwave filter pair, an amplitude comparison function (i.e., a ratio between the two transfer functions of the filter pair) is obtained, that is, quasi-linear and monotonically decreasing over a large RF frequency band. Then, the measurement of the microwave frequency can be done by simply measuring the microwave powers from the two outputs of the photonic microwave filter pair. A microwave frequency measurement range as large as 36 GHz has been demonstrated, with a measurement accuracy better than ±200 MHz.

The solution proposed in [21] exploits the concept of a channelized receiver, associated with the time-to-frequency mapping to allow a more precise measure of the instantaneous frequency by applying a Fourier analysis to the time-to-frequency transformed signal. In practice, an optical ultra-short pulse is first dispersed in a highly dispersive element (e.g., a spool of dispersion compensating fiber). The dispersed pulse becomes a long pulse, whose spectral components are distributed in time from the lowest to the highest (or vice versa, depending on the sign of the dispersion). Then, the dispersed pulse is modulated by the received microwave signal, encoding the different time instant into a different optical wavelength. A channelizer follows, which sends different portions of the spectrum to different low-BW PDs and ADCs, organized in an array. The powers at the outputs of the channels are measured and constitute the sampled version of the microwave signal under test. Finally, the gathered data are analyzed using a digital processor, and the spectral distribution of the temporal signal is obtained in real time from a Fourier transform analysis. In the cited work, the measurement resolution is estimated to be 300 MHz, depending on the temporal duration of the pulse (3 ns), and the frequency measurement error is about 30 MHz. This approach shows the main advantage of allowing simultaneously measuring different frequency components of the signal.

7.5 Scanning receivers

As seen in the previous paragraph, the channelized receivers proposed so far, although potentially simple, do not allow to precisely determining the frequency of the unknown RF signals. A better performance can be reached implementing a scanning receiver instead.

The basic scheme of a scanning receiver is reported in Figure 7.5. The principle is based on a tunable and narrowband filter that scans the modulation sideband of a laser carrying the RF spectrum, so that the optical power detected by a PD can be associated to the position of the filter, that is, to the analyzed RF frequency.

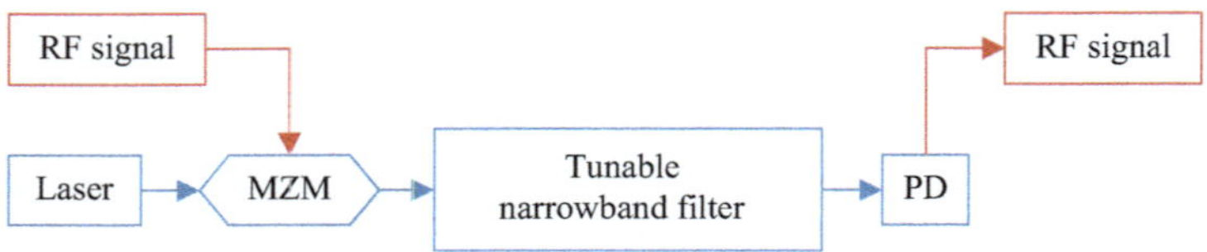

Figure 7.5 Scheme of a scanning receiver based on a tunable narrowband filter

This approach was proposed back in 1999 in [22], where a scanning receiver was already demonstrated with a BW of 40 GHz. The system used the temporal behavior of a piezoelectrically scanned fiber Fabry–Perot etalon, with an RF resolution of 90 MHz across the entire scan range. Due to the filter and its tuning mechanism, the scan time was of 15 ms, therefore obtaining only a low probability of intercept. Nevertheless, the scheme demonstrated the potentials of this approach: the extremely wide scan range, the low BW of the post-processing electronics, and the compatibility with optical-fiber antenna remoting. On the other hand, besides being slow, it suffered limited sensitivity and SFDR, and was sensitive to frequency drift in the laser and filter.

More recently [23], the tunable narrowband filter has been realized as a fiber Bragg grating (FBG) written in a microstructured fiber with internal electrodes, which has showed a sweep time as short as a few hundred us. The filter's frequency sweep has been achieved by heating the internal electrodes of the fiber with a short electrical pulse that shifts the Bragg wavelength of the grating to longer wavelengths. After the current pulse, the fiber cools back to the initial temperature.

The used filter has showed a transmission peak with a BW of 54 MHz, and a rejection of 33 dB at 2 GHz from the peak. Although the current pulse shifts the filter wavelength very quickly, the cooling of the FBG limits the repetition rate of the pulse generator to 200 Hz.

The examples of scanning receivers reported above show several limitations that dampen their practical use: the scanning speed is usually slow (in the order of few hundred Hz), the resolution is limited to the BW of the tunable filter (tens of MHz), and they do not resolve the time behavior of the signals. In the next paragraphs, we report on a recently proposed approach of scanning receiver that overcome all these limitations, showing a precise frequency measure resolution and a high dynamic range, with the capability of analyzing the time behavior of RF signals and the potential for a fast scanning time.

7.6 A photonics-based coherent scanning receiver

In EW applications, the ideal receiver would be a wideband software defined receiver, able to replace all the hardware functionalities with advanced digital signal processing (DSP) techniques [24]. Albeit the concept and expectations of software defined receivers are continuously evolving, digital electronics does not evolve so quickly and the current practical solution for EW receivers is still realized through a channelized implementation. In the channelized receiver, the RF spectrum is simultaneously detected by several heterodyne receivers, down-converting adjacent portions of the spectrum to a fixed intermediate frequency (IF). Then, a set of ADCs simultaneously acquire all the spectrum portions providing a complete picture of the electromagnetic environment. Although this approach reaches high performance, it is characterized by large size, weight, power consumption (SWaP), and requires a huge effort in the design of the filter banks.

An interesting alternative solution to the channelized receiver would be scanning the spectrum through a single tunable heterodyne receiver and a single ADC, thus reducing the SWaP and increasing the agility. Unfortunately, current LOs and image rejection filters suffer very limited tuning capability, spoiling the applicability of the scanning receiver [25]. On the other hand, direct conversion receivers avoid the use of image reject filters, exploiting instead LPFs, resulting in a reduced complexity with respect to the heterodyne architecture [26]. Nevertheless, the mixers induce flicker noise and second-order intermodulation distortion (IMD2), and they show finite port-to-port isolation (LO-to-RF leakage and RF-to-IF feedthrough), resulting in detrimental DC offset and spurious signals at baseband. In more detail, the DC offset is a particularly serious drawback because it can saturate the high gain amplification stages that follow the mixer [27]. Several methods have been introduced to reduce the DC offset value to the order of few mV, such as using sub-harmonic mixers [26], active AC coupling [28], feedback tuning loops [27], etc., which often vanish the straightforwardness of the direct conversion receiving architecture. Moreover, the tunability issues of the LOs still remain.

As reported above and in Chapter 3, photonic technologies are recently demonstrating attractive features for microwave applications including EW applications, as ultra-wide BW, tunable filtering [29], photonics-based microwave mixing with very high port-to-port isolation [30], and intrinsic immunity to electromagnetic interferences. Therefore, photonic techniques are being proposed in RF receivers to implement the filtering and down-conversion tasks over wide frequency bands [31].

In [32,33], we have proposed a photonics-based direct conversion receiver that scans in discrete steps the RF spectrum, which is immune to the issues of LO self-mixing and RF-to-baseband feedthrough.

Figure 7.6 reports the scheme of principle. The proposed architecture transfers the detected RF spectrum to the optical domain and exploits an optical LO to convert a selectable portion of the detected spectrum back to baseband to be digitally detected and analyzed. Changing the detuning between the converted RF spectrum and the optical LO, the whole RF spectrum is scanned step by step by using a single digital detector. The scheme therefore implements a coherent scanning receiver avoiding the need of tunable optical filters or high-frequency RF LOs, and allows a fast scanning of the detected spectrum. Clearly, a phase locking between the RF spectrum in the optical domain and the LO is necessary in order to preserve the RF input signal integrity, otherwise phase noise between them dramatically corrupts the detected signal [34].

Figure 7.7 shows the scheme in more detail. The RF-to-optical conversion is realized modulating a tunable master laser (ML) with the detected RF spectrum in an MZM. The ML is a distributed-feedback laser (DFB) at optical frequency ν_{ML} in

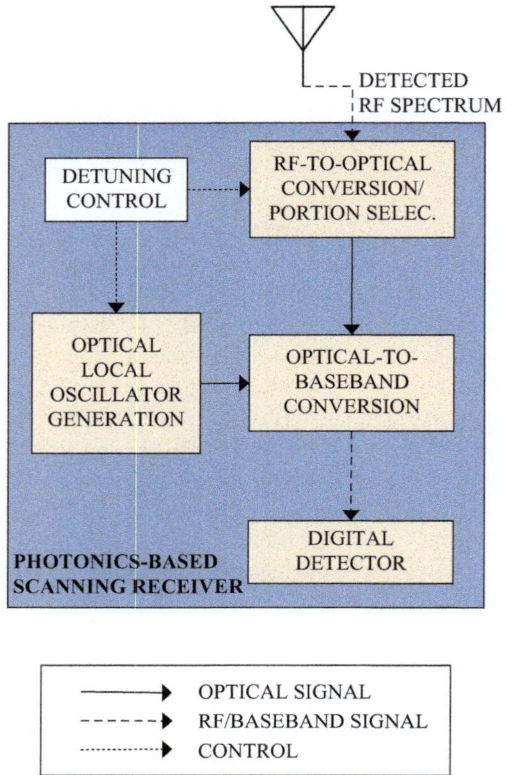

Figure 7.6 Principle of operation of the photonics-based coherent scanning receiver

the communication band around 1550 nm (inset i in Figure 7.7). The MZM is set to realize a double-sideband modulation with carrier suppression (inset ii). The obtained optical spectrum is coupled in an I/Q demodulator with a slave laser (SL) at frequency ν_{SL}, acting as the optical LO (inset iii). The I/Q demodulator is composed of a 90° optical HC and two BPDs. Thanks to the beatings at the BPDs between the modulation sideband and the SL, the I/Q demodulator realizes the optical-to-baseband conversion of the I and Q components of the modulation sideband, centered at the wavelength of the SL. The HC combines the signal of the modulated ML with the four quadrature states of the LO (which can be expressed as ML + SL, ML − SL, ML + jSL, ML − jSL), so that the BPDs can generate an electrical signal proportional to the I and Q components of the down-converted RF spectrum. Then, two LPFs pick out the baseband portion of the I and Q components up to 500 MHz (inset *iv*). Finally, the filtered components are digitized by two ADCs at 2 GSample/s, and a DSP calculates the 1GHz-BW complex envelope (inset v). The scan of the entire detected RF spectrum is performed tuning the ML step by step, while keeping the SL at a fixed wavelength. At every tuning step of

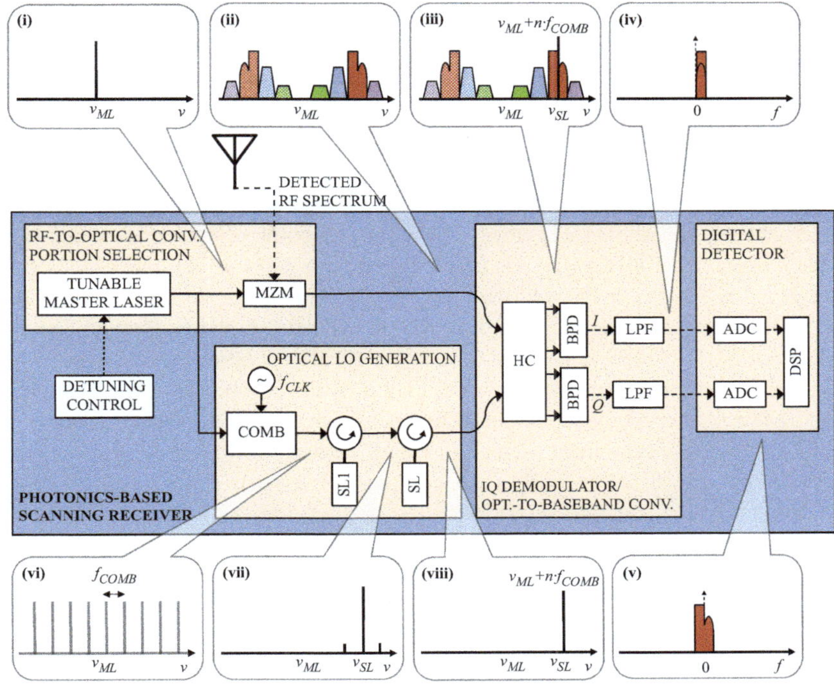

Figure 7.7 The scheme of the photonics-based coherent scanning receiver. Solid lines indicate optical connections, while dashed lines indicate RF or baseband connections. Dotted lines are for controls. For the sake of clarity, the scheme does not show the Er-doped fiber amplifier boosting the ML, and the electrical amplifiers before the ADCs. The insets (i)–(viii) sketch the optical and electrical signals along the scheme. The optical frequencies are indicated as ν, *while the RF/baseband frequencies are denoted as* f. *At the nth tuning step, the SL is locked to the nth comb mode, the detuning between ML and SL is* $n{\cdot}f_{CLK}$, *and the photonics-based scanning receiver down-converts to baseband the portion of the detected RF spectrum at frequency* $n \cdot f_{CLK}$

the ML, the modulation sideband is shifted with respect to the SL, and a different portion of the detected RF spectrum is down-converted to baseband. The scheme therefore realizes a so-called tunable direct-conversion receiver.

7.6.1 Architecture of the photonics-based coherent scanning receiver

As said above, a phase locking between the master and SLs is necessary. The proposed phase locking technique exploits the generation of a laser comb from the ML and injects the comb in the SL so that the SL locks its frequency and phase to the comb line aligned to its resonance frequency. Consequently, the SL locks to the ML as well. At every tuning step of the ML, the comb is shifted to align the next comb mode to the SL, injecting and locking it. The detuning between the ML and the SL defines the RF spectrum portion to be down-converted and analyzed. The comb is generated from the ML through a cascade of intensity and phase modulators [31] driven by a clock tone at $f_{CLK} = 3$ GHz. The obtained comb with 3 GHz spacing is then densified by modulating it in an additional MZM driven by a tone at $f_{COMB} = 1$ GHz derived dividing the 3 GHz clock. The result is a comb of optical lines at 1 GHz from each other (inset vi). The ML is tuned in steps of 1 GHz so that one of the comb lines falls at the resonance frequency of the SL. The optical comb injects a first DFB laser (SL1) at optical frequency ν_{SL} through an optical circulator. Besides locking the phase of SL1 to that of the ML, the injection also suppresses the other comb lines, resulting in the selection and amplification of the comb line at frequency ν_{SL} (inset vii). To further increase the comb line selection, that is, the spectral purity of the SL, SL1 injects a second SL, which is therefore phase-locked to the ML as well, and acts as the optical LO at the I/Q demodulator (inset viii).

7.6.2 Features and figures of merits

Figure 7.8 shows a picture of the receiver demonstrator, covering the input frequency range from 500 MHz to 40 GHz. The box in the picture contains all the photonic components, including the related controls and the power supplies, while the ADCs and the DSP are left outside.

Figure 7.8 Picture of the photonics-based scanning receiver

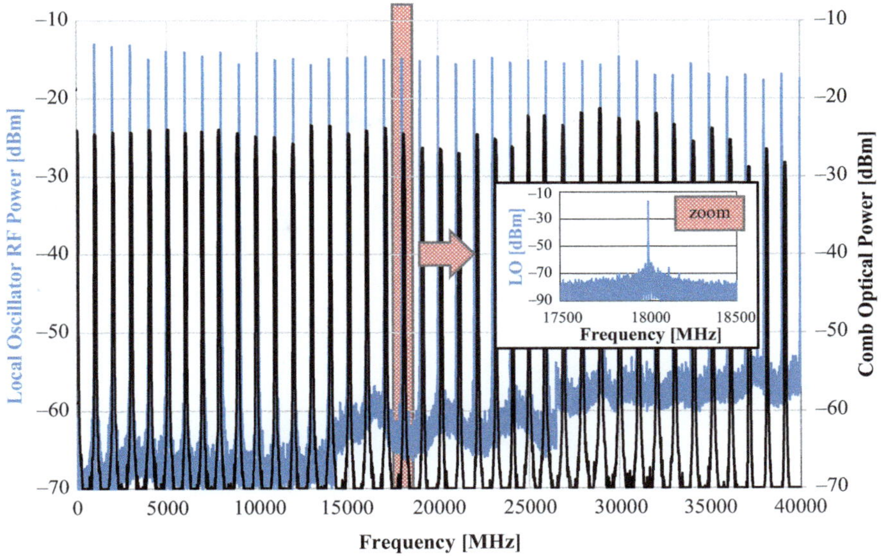

Figure 7.9 Effect of the injection locking operation. Black trace: optical spectrum of the laser comb. Blue trace: electrical spectrum of the beating between the injected SL and the tunable ML in a PD, for each of the tuning positions. Inset: zoom trace of the electrical spectrum of the LO at 18 GHz

Figure 7.9 shows the performance of the injection-locking operation. The black trace reports the optical spectrum of the laser comb, while the blue trace shows the injected SL at each tuning position, as detected by an electrical spectrum analyzer (with 20 MHz resolution BW) after beating with the ML. The black trace shows the upper half of the optical comb, with the line at frequency detuning of 0 MHz being the ML. The inset shows the zoom trace of the electrical spectrum of the LO at 18 GHz (with 300 kHz resolution BW). More than 40 lines from the comb succeed in stably locking the SL, allowing the down-conversion of the whole RF spectrum from 0.5 to 40.5 GHz, through contiguous channels of 1 GHz width. A locking range of about 200 MHz has been measured for the SL, relaxing the requested precision in tuning the ML. Moreover, the comb lines stably lock the SL within an injected optical power range of about 20 dB, with minimal output power variations (<1 dB).

Figure 7.10 shows an example of scanning the RF spectrum, here limited between 0.5 and 18.5 GHz, with the photonics-based receiver [35]. The signals emitted by four lab-grade RF generators are used to emulate an electromagnetic scenario crowded with typical signals for both surveillance and communications. The receiver scans the RF spectrum and reconstructs the detected signal merging the down-converted outputs from each tuning step. The graph shows the power spectral density of the signal calculated from the fast Fourier transform (FFT). The four signals are clearly detected in terms of their mean power. Their time-varying

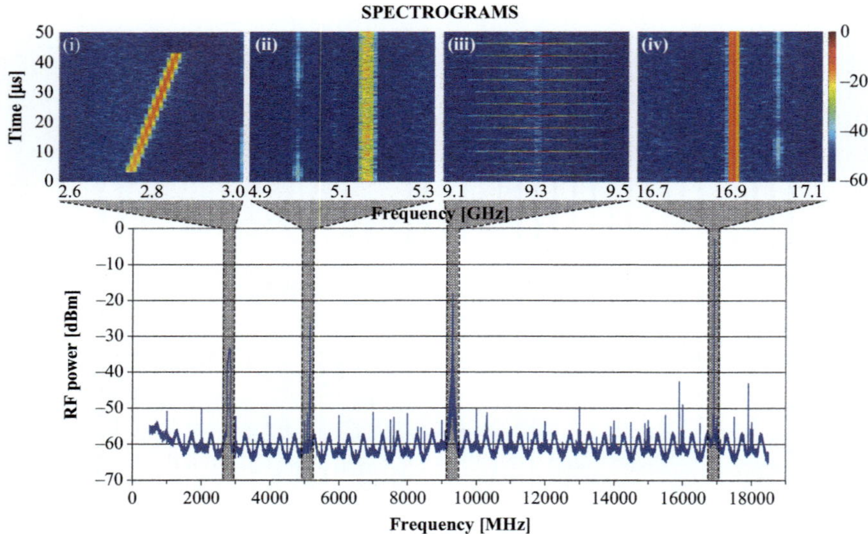

Figure 7.10 Example of a full spectrum scan with the photonics-based scanning receiver. The insets (i)–(iv) report the corresponding spectrograms related to the four detected signals

nature is analyzed through spectrograms, calculating the FFT for successive short acquisition periods. The insets in Figure 7.10 report the spectrograms obtained considering a typical FFT length for electronic support measures (ESM) receivers equal to 256 samples only (corresponding to 128 ns) [36]. It is possible to recognize that the signal at 2.82 GHz is a pulse with a linear frequency chirp of 100 MHz and a mean power of –34 dBm (inset i), the signal at 5.15 GHz is a signal with 20 MHz BW (it is actually a Wi-Fi signal, 802.11.n) and –27 dBm mean power (inset ii), the signal at 9.30 GHz is a pulsed signal with very short pulses of few μs (inset iii) and –18 dBm mean power, while the signal at 16.9 GHz is a continuous-wave (CW) tone with –3 dBm mean power (inset iv). At each tuning step of the ML, the input spectrum has been acquired for 500 μs, corresponding to the maximum memory depth of the exploited ADCs. The reconstructed spectrum in Figure 7.10 also shows few other spurious tones that are all suppressed more than 40 dB with respect to the highest input signal, as requested by the ESM systems, confirming the good sensitivity and dynamic range of the architecture. While the tones at 15.9 and 17.9 GHz are due to the crosstalk induced by the strong CW signal at 16.9 GHz on the two adjacent channels, the tones at multiples of 1 GHz are caused by the DC offset of the ADCs' amplification stages and can be electronically compensated for, further improving the receiver performance.

A complete laboratory characterization of the photonics-based coherent scanning receiver (see Table 7.2) has proven performance on par with state-of-the-art channelized receivers [35], despite its early demonstrator stage, thanks to the effectiveness of the photonics-based architecture enabling the tunable direct conversion.

Table 7.2 Performance of the photonics-based scanning receiver, compared with state-of-the-art channelized receivers

Parameter	Photonics-based scanning receiver demonstrator	Channelized receiver-state of the art
Input RF frequency range	0.5 GHz ÷ >40 GHz	0.5 GHz ÷ 18 GHz
Input RF power range[a]	–80 dBm ÷ 0 dBm	–80 dBm ÷ 0 dBm
RF instantaneous 3 dB BW	1 GHz	1 GHz
RF instantaneous dynamic range[b]	> 40 dB	> 40 dB
Input third-order intercept point (IIP3)	28 dBm	> 25 dBm
Input second-order intercept point (IIP2)	57 dBm	> 35 dBm
Settling and tuning time[c]	Can be < 100 ns	< 100 ns
Noise figure (NF)[d]	< 10 dB	10 dB
Dimensions[e]	6U, thickness < 100 mm	6U, thickness < 35 mm, per channel

Notes:
[a]The input power range extends from the input RF power that ensures an output SNR > 8 dB (considering a resolution BW of 40 MHz and a preamplifier stage) to the 1 dB compression point.
[b]The instantaneous dynamic range is defined as the ratio between the input RF signal power and the strongest spurious tone generated by the receiving system, neglecting harmonics and intermodulation products.
[c]The settling and tuning time of the photonics-based scanning receiver is already as low as few us, but an ML with a phase-control section can avoid the thermal fluctuations and push it to < 100 ns.
[d]The reported NF is reached considering a preamplifier stage (with gain 30 dB and NF 4 dB). Without preamplifier, the photonics-based scanning receiver shows a NF < 40 dB.
[e]The dimensions of the photonics-based scanning receiver can be further reduced using photonic integration. A single 6U module with few centimeter thickness is expected for a 6-channel version.

The measured instantaneous dynamic range (>40 dB) is sufficiently high to guarantee the correct behavior of the receiver in ESM applications (see requirements reported in Chapter 2) while an RF dynamicator (usually exploited in current ESM receivers as well) allows using the full dynamic range of 80 dB. Moreover, the second-order intermodulation distortion is particularly low (IIP2 at 57 dBm) thanks to the odd frequency response of the MZM. Finally, differently from electronics-based direct conversion receivers, the exploited photonic down-conversion is immune to RF-to-baseband feedthrough and a residual DC lower than 50 μV has been measured.

The RF scanner is tuned by controlling the driving current of the ML, rapidly changing its cavity refractive index. This tuning method is potentially very fast as it involves the plasma effect due to carrier variation that shows response time below 1 ns. Unfortunately, it also induces thermal transients due to the change in current absorption on the laser junction, which slow down the laser response. Due to the thermal transients, in the current implementation we have obtained a channel selection time of ~10 ms. Anyway, few solutions are available to meet the fast switching time requirement of a ESM RF scanner. For example, one is using an ML that avoids the thermal transients, as, for example, a distributed Bragg reflector laser (which incorporates gain, phase, and grating sections) that uses the quantum-confined Stark effect to achieve fast tuning response without changes in the absorbed current. A similar laser has been already demonstrated allowing a tuning time <1 ns [37].

The developed scanning receiver is currently implemented as a benchtop box (which also includes the power supplies and controls, see Figure 7.1), and it is already engineered as a single 6U Eurocard although it makes use of commercial discrete components only, setting a significant size reduction with respect to ESM channelized receivers. A system implementation through photonic integration techniques, which is under consideration, will guarantee even smaller size and weight, allowing the use of ESM receivers in the most demanding applications. Moreover, the proposed architecture allows easily remoting the antenna from the receiver, moving the MZM to the antenna site through optical fibers, with negligible additional weight and losses with respect to the presented solution, and setting another strong advantage over current channelized receiver that must either be put at the antenna site, or use heavy and cumbersome RF waveguides.

7.7 Case study: a field trial in a tactical naval scenario

In a real scenario, the threat radars use complex low-probability-of-intercept waveforms (as ultra-wideband signals and frequency hopping) to confuse the ESM receivers so that they cannot jam the threat. Therefore, we have organized a field-trial experiment to test the receiver in a relevant outdoor environment. The field trial has been run at the Center for Support and Naval Experimentation – Institute for Telecommunications and Electronics (CSSN-ITE) of the Italian Navy in Livorno, using typical operative signals composed of multiple pulses with agility in modulation, center frequency, pulse width (PW), and pulse repetition interval (PRI) [35].

Figure 7.11 shows the satellite view of the field trial area. The transmitting system has been implemented using a scenario generator emulating typical threats, and it has been settled on a mobile lab (inset A) provided with a 30 dBi gain wideband parabolic antenna. The photonics-based coherent scanning receiver has been set 250 m apart (inset B), collecting the signal with a 10 dBi gain horn antenna placed on the roof of the building (inset C), and amplifying it by a 30 dB gain wideband amplifier.

Figure 7.12 reports the spectrograms obtained with two typical scenarios. In Figure 7.12(a), a pulsed signal with central frequency changing between 8.91 and 9.08 GHz and a peak power of –10 dBm is received. The pulses have a PW of 0.8 μs and a PRI of 8.6 μs, and they are organized in bursts of six pulses (three at each frequency), with a burst period of 103 μs. The high pulse power allows identifying the image signals, with an ER of 41 dB. On the other hand, Figure 7.12(b) reports a low-duty cycle pulsed signal at 10.1 GHz summed to two widely chirped pulses at frequencies 9.97 and 10.1 GHz. The first chirped pulse has a PW of 60 μs and a PRI of 410 μs, an initial frequency of 10.1 GHz and a linear chirp across a BW of

Figure 7.11 The field trial site at CSSN-ITE in Livorno, Italy: (a) Transmitting system on the mobile lab; (b) receiver site (the circle indicates the demonstrator); (c) receiving horn antenna

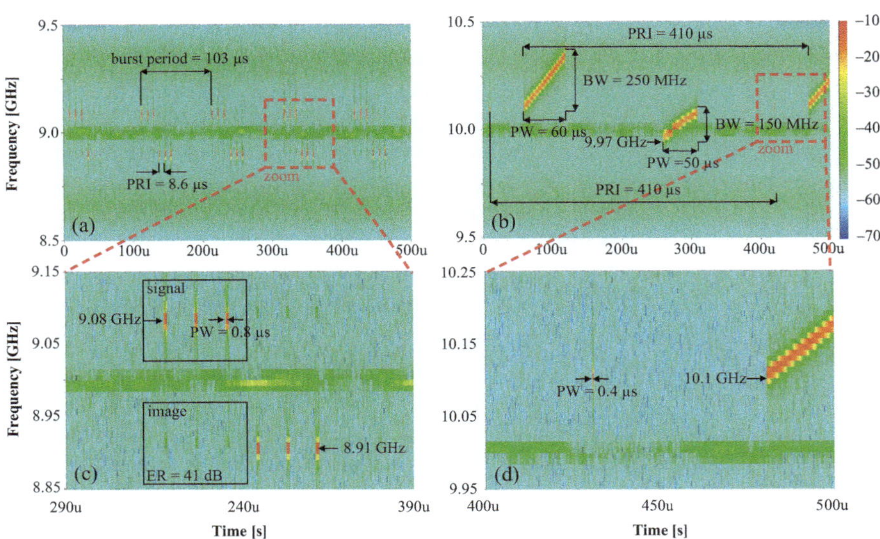

Figure 7.12 (a, b) Spectrograms acquired during the field trial at the CSSN-ITE, emulating typical ESM scenarios; (c, d) corresponding detailed zooms of the acquisitions

250 MHz. The second chirped pulse has a PW of 50 μs, an initial frequency of 9.97 GHz, and a BW of 150 MHz. The PRI cannot be measured because only one pulse has fallen into the observation window of 500 μs. The low-duty cycle pulse has a PW of 0.4 μs at 10.1 GHz and a PRI of 410 μs. The generated waveforms have been scaled down in time, with respect to common radar threats, to match the maximum acquisition window of the exploited ADCs. Detailed zooms of the spectrograms are

shown in Figure 7.12(c) and (d), respectively. From the graphs, the received signals can be clearly identified and their parameters (PW, PRI, central frequency, etc.) can be verified. In particular, in the first scenario the receiver demonstrator correctly detects the pulsed signal with a frequency hopping occurring every three pulses in bursts of six. The strong pulses show a peak power of −10 dBm that allow identifying image signals with a rejection of 41 dB, confirming the previous lab characterization in terms of dynamic range. In the second scenario, the two linearly chirped pulse trains are clearly recognizable. The low-duty-cycle pulse train is clearly visible as well. The field trial therefore has confirmed the receiver performance in terms of high linearity and sensitivity also in a relevant outdoor scenario.

7.8 Summary

In this chapter, we have described several applications of photonics to electronic warfare systems. The reported results, the possibility to combine the schemes with the antenna remoting, and the photonics robustness to electromagnetic interferences substantiate the suitability of the photonics-based approaches for next ESM applications, in particular for the coherent scanning receiver. Nevertheless, it is also fundamental for high-demanding EW application to reduce significantly the size of the systems, their weight and power consumption (SWaP). The rapid maturity of photonic integration technologies and 3D hybrid packaging suggest having the potential also for reaching this additional requirement. It is important to note here that the time and costs for designing, realizing, and fully developing these photonic-integrated EW systems require a direct industrial interest and investment. Actually, industries are observing with evident interest the development of the photonic technology in this field. Although they still are not fully taking the lead, the increasing amount of demonstrations with very competitive results are pushing industries to jump in and accelerate the development process, opening the way to new applications in civil fields as well.

References

[1] K. Davis, J. Gray, and A. Stark, “Photonics components and subsystems for electronic warfare,” *Avionics and Vehicle Fiber-Optics and Photonics Conference*, AVFOP, San Diego, CA, USA, 2013.

[2] J. Capmany, B. Ortega, and D. Pastor, “A tutorial on microwave photonic filters,” *IEEE J. Lightw. Technol.*, vol. 24, no. 1, pp. 201–229, 2006.

[3] A.J. Stark, K. Davis, C. Ward, and J. Gray, “Photonics for electronic warfare,” *Avionics and Vehicle Fiber-Optics and Photonics Conference*, AVFOP, Atlanta, GA, USA, 2014.

[4] C.H. Cox, and E.I. Ackerman, “Photonics for simultaneous transmit and receive,” *Microwave Symposium Digest (MTT), 2011 IEEE MTT-S International*, no., pp.1, 4, 5–10 June 2011.

[5] M.E. Manka, "Microwave photonics for electronic warfare applications," International Topical Meeting on Microwave Photonics, MWP, 2008.

[6] T.E. Darcie and J. Zhang, "High-performance Microwave-photonic links," *Radio and Wireless Symposium*, RWS, 2008.

[7] V. Urick, M. Rogge, P. Knapp, L. Swingen, and F. Bucholtz, "Wide-band predistortion linearization for externally modulated long-haul analog fiber-optic links," *IEEE Trans. Microw. Theory Tech.*, vol. 54, no. 4, pp. 1458–1463, 2006.

[8] T. Darcie and P. Driessen, "Class-AB techniques for high-dynamic-range microwave-photonic links," *IEEE Photon. Technol. Lett.*, vol. 18, no. 8, pp. 929–931, 2006.

[9] S. Mathai, F. Cappelluti, T. Jung, *et al.*, "Experimental demonstration of a balanced electroabsorption modulated microwave photonic link," *IEEE Trans. Microw. Theory Techn.*, vol. 49, no. 10, pp. 1956–1961, 2001.

[10] C. Cox, III, E. Ackerman, G. Betts, and J. Prince, "Limits on the performance of rf-over-fiber links and their impact on device design," *IEEE Trans. Microw. Theory Tech.*, vol. 54, no. 2, pp. 906–920, 2006.

[11] V. Urick, M. Rogge, F. Bucholtz, and K. Williams, "The performance of analog photonic links employing highly compressed erbium-doped fiber amplifiers," *IEEE Trans. Microw. Theory Tech.*, vol. 54, no. 7, pp. 3141–3145, 2006.

[12] C. Lim, M. Attygalle, A. Nirmalathas, D. Novak, and R. Waterhouse, "Analysis of optical carrier-to-sideband ratio for improving transmission performance in fiber-radio links," *IEEE Trans. Microw. Theory Tech.*, vol. 54, no. 5, pp. 2181–2187, 2006.

[13] M. Attygalle, K. Gupta, and T. Priest, "Broadband extended dynamic range analog signal transmission through switched dual photonic link architecture," *IEEE Phot. J.*, vol. 3, no. 1, pp. 100–111, 2011.

[14] R. Kalman, J. Fan, and L. Kazovsky, "Dynamic range of coherent analog fiber-optic links," *J. Lightw. Technol.*, vol. 12, no. 7, pp. 1263–1276, 1994.

[15] T.R. Clark, S.R. O'Connor, and M.L. Dennis, "A phase-modulation I/Q-demodulation microwave-to-digital photonic link," *IEEE Trans. Microw. Theory Techn.*, vol. 58, no. 11, pp. 3039–3058, 2010.

[16] A. Ramaswamy, L. Johansson, J. Klamkin, *et al.*, "Integrated coherent receivers for high-linearity microwave photonic links," *J. Lightw. Technol.*, vol. 26, no. 1, pp. 209–216, 2008.

[17] Y. Li, W. Jemison, P. Herczfeld, and A. Rosen, "Coherent, phase modulated (PM) fiber-optic link design," *2006 IEEE MTT-S International Microwave Symposium Digest*, San Francisco, CA, USA, 2006.

[18] D.B. Hunter, L.G. Edvell, and M.A. Englund, "Wideband microwave photonic channelised receiver," International Topical Meting Microwave Photonics, 2005.

[19] Z. Li, X. Zhang, H. Chi, S. Zheng, X. Jin, and J. Yao, "A reconfigurable microwave photonic channelized receiver based on dense wavelength

division multiplexing using an optical comb," *Optics Comm.*, no. 285, pp. 2311–2315, 2012.

[20] S. Pan and J. Yao, "Instantaneous microwave frequency measurement using a photonic microwave filter pair," *IEEE Photon. Technol. Lett.*, vol. 22, no. 19, pp. 1437–1439, 2010.

[21] C. Wang and J. Yao, "Ultrahigh-resolution photonic-assisted microwave frequency identification based on temporal channelization," *IEEE Trans. Microw. Theory Techn.*, vol. 61, no. 12, pp. 4275–4282, 2013.

[22] S.T. Winnall and A.C. Lindsay, "A Fabry–Perot scanning receiver for microwave signal processing," *IEEE Trans. Microw. Theory Techn.*, vol. 47, no. 7, pp. 1385–1390, 1999.

[23] P. Rugeland, Z. Yu, C. Sterner, O. Tarasenko, G. Tengstrand, and W. Margulis, "Photonic scanning receiver using an electrically tuned fiber Bragg grating," *Opt. Lett.*, vol. 34, no. 24, pp. 3794–3796, 2009.

[24] C. Dantea, *Modern Communications Receiver Design and Technology*, Artech House Publishers, August 2010.

[25] W. Ruijia and W. Xing "Radar emitter recognition in airborne RWR/ESM based on improved K nearest neighbor algorithm," *2014 IEEE International Conference on Computer and Information Technology*, Xi'an, China, 2014, DOI 10.1109/CIT.2014.115.

[26] I. Arasaratnam, S. Haykin, T. Kirubarajan, and F. A. Dilkes "Tracking the mode of operation of multi-function radars," *2006 IEEE Conference on Radar*, Verona, NY, USA, 2006, DOI: 10.1109/RADAR.2006.1631804.

[27] A. Wesley, "Radar warning receiver (RWR) time-coincident pulse data extraction and processing," *IEEE Radar Conference*, Atlanta, GA, USA, 2012.

[28] J. Coward, "Analog to information (A2I) sensing for software defined receivers," Navy SBIR FY2009.2, 2009.

[29] K.-Y. Tu, M.S. Rasras, D.M. Gill, *et al.*, "Silicon RF-photonic filter and down-converter," *J. Lightw. Technol.*, vol. 28, no. 20, pp. 3019–3028, 2010.

[30] A.C. Lindsay, G.A. Knight, and S.T. Winnall, "Photonic mixers for wide bandwidth RF receiver applications," *IEEE Trans. Microw. Theory Tech.*, vol. 43, no. 9, pp. 2311–2317, 1995.

[31] X. Xie, Y. Dai, K. Xu, *et al.*, "Broadband photonic RF channelization based on coherent optical frequency combs and I/Q demodulators," *IEEE Photon. J.*, vol. 4, no. 4, pp. 1196–1202, 2012.

[32] D.Onori, F. Scotti, F. Laghezza, *et al.*, "A photonically enabled compact 0.5–28.5 GHz RF scanning receiver," *J. Lightw. Technol.*, vol. 36, no. 10, pp. 1831–1839, 2018.

[33] D. Onori, F. Scotti, F. Laghezza, *et al.*, "0.5–40 GHz range extension of a compact electronic support measures scanning receiver based on photonics," *18th International Radar Symposium IRS 2017*, Prague, Czech Republic, 2017.

[34] R.T. Logan, Jr, "Photonic radio frequency synthesizer," *SPIE's 1996 International Symposium on Optical Science, Engineering, and Instrumentation*, Denver, CO, USA, 1996.

[35] D. Onori, F. Scotti, F. Laghezza, *et al.*, "Relevant field trial of a photonics-based RF scanning receiver for electronic support measures," TuM2.3, 2016 IEEE International Topical Meeting on Microwave Photonics (MWP), 2016.

[36] R.G. Wiley, *ELINT: The Interception and Analysis of Radar Signals*, Artech House Radar Library, 2006.

[37] M. Pantouvaki, C.C. Renaud, P. Cannard, M.J. Robertson, R. Gwilliam, and A.J. Seeds, "Fast tuneable InGaAsP DBR laser using quantumconfined stark-effect-induced refractive index change," *IEEE J. Select. Topics Quantum Electron.*, vol. 13, no. 5, pp. 1112–1121, 2007.

Chapter 8

Past and future of radars and EW systems: an industrial perspective

Alfonso Farina[1]

8.1 Chapter organization and key points

This chapter focuses on the past, present, and future of photonics technology in RAdio Detection And Ranging (RADAR) and electronic support measures (ESM) systems. With respect to what has been presented in the other chapters, here we add comments from an industrial point of view on few additional functions enabled by photonics, namely the optical processors and the quantum radars.

We start with "the operational needs" of wide (may be as wide as a country territory) surveillance critical infrastructures. These will include the dual-use surveillance too: for example, the humanitarian needs related to the immigration across the Mediterranean Sea.

Due to the relevance of these operational needs, the interdisciplinary nature of the technical approach has to be encouraged. The collaboration should be fostered between industry, university, governmental and legal organizations, and the *users* (indeed all the stakeholders involved in the management of these complex critical infrastructures), which contribute to establish and refine the operational needs.

Radar and electronic warfare (EW) systems can operate on separate platforms, as well as they can coexist on just one platform like a naval ship. The coexistence of radar and EW systems is a challenge from the technology, technical, engineering, and systemic points of view. This problem should be carefully tackled, and it will be briefly afforded in Section 8.7.

The chapter is structured as follows. The operational needs are described in Section 8.2. The role of photonics in surveillance phased array radar (PAR) is afforded in Section 8.3. Synthetic aperture radar (SAR) and the role played by optronics is the theme of Section 8.4. How optronics could greatly help in the adaptive signal processing for digital beam forming in surveillance radar and SAR modes is discussed in Section 8.5. The role of optronics in ESM is dealt with in Section 8.6. The coexistence of radar and EW is the theme of Section 8.7. Quantum sensing and quantum radar is afforded in Section 8.8: the question that

[1]Selex Sistemi Integrati, Italy, retired

is asked is: Sci-Fi or a potential reality? The chapter closes with a summary and way-ahead ideas. Then, references quoted in the text and additional references of potential interest to the reader but not quoted in the text are included.

8.2 Operational needs

Operational needs are those statements that "identify the essential capabilities, associated requirements, performance measures, and the process or series of actions to be taken in effecting the results that are desired in order to address mission area deficiencies, evolving applications or threats, emerging technologies, or system cost improvements." The operational requirements assessment starts with the Concept of Operations and goes to a greater level of detail in identifying mission performance assumptions and constraints and current deficiencies or enhancements needed for operations and mission success. Operational requirements are the basis for system requirements [1].

8.2.1 Radar

Modern surveillance systems (sensors, command and control, reaction) have to comply with an unprecedented wide spread of operational requests. Here is a not exhaustive list:

- Surveillance/classification everywhere, every time, every domain, with guaranteed low level of false alarm rate and, hopefully, in the respect of privacy and national–international regulations.
- Take care of many types of platforms (underwater, ground based, shipborne, airborne, unmanned vehicles, drones, spaceborne) carrying sensors for a wide spectrum of applications:
    - Civilian (air traffic control (ATC) and weather forecast and monitoring), humanitarian (immigration/people migration).
    - Defense (against fast-moving fast-maneuvering low radar cross section (RCS) targets, far-range tactical ballistic missile since boost phase, from a wide spread of rocket types, from mortar shots, from drones), non-cooperative target recognition (NCTR), electronic attack (EA).
    - Security, underground inspection, through the wall surveillance (as reported by the VisiBuilding program [2]).
    - Airborne/satellite to monitor quality of life in the planet (weather forecast also at short time, oceanography, estimation of space–time wind field also for exploitation in the windfarm management, estimation of bio mass, pollution, etc.).
    - Surveillance/tracking/classification and mitigation of space debris.
- Localization of our own forces/resources.
- Go beyond the extraction of usual information to get intelligence from the measurements.
- Go beyond the usual way to show data and get wide/wise dissemination complying with security laws and cybersecurity.

- Operate in a limited electromagnetic (e.m.) spectrum resource (which today is a commodity) also complying with international and security regulations.
- Cooperate, coexist with other e.m. apparatuses in the operational field.
- Coordinate and operate sensor management online/on the fly for joint improved performance.

Radar and EW are the key sensors we will refer mainly to it in this chapter. Radar goes arm-in-arm with clutter, that is, the e.m. reverberation from the surrounding environment. From a radar engineer point of view, clutter is like a two-faced Janus: the worst enemy and at the same time the best friend. In fact, clutter is always present when the radar is switched on and its deleterious effects are to be mitigated to detect the target we are looking for (a noticeable exception being the SAR where the clutter is indeed the spatially widespread target to image). At the same time, without the need to contrast clutter, radar engineers would have much less work to do and minor revenues from their work.

Modern radar are more powerful than ever because they look for dim targets, thus the amount of clutter to manage is sometimes very large. Improving radar sensitivity for target detection implies more powerful clutter mitigation techniques. From a hardware point of view, there are two main practical limitations for clutter rejection: dynamic range and reference oscillator phase noise. Technology plays a key role in these fields. Signal processing algorithms do the clutter mitigation, target extraction, parameter estimation, NCTR, situational awareness, etc.

8.2.2 EW

EW is defined as a military action involving the use of EM energy to determine, exploit, reduce, or prevent radar use of the EM spectrum. The operational employment of EW relies upon the capture of radar EM emissions using electronic intelligence devices, collating the information in support databases that are then used to interpret EM emission data, to understand the radar and other e.m. system functions, and to program reactions against these systems. EW is organized into two major categories: electronic warfare support measures (ESM) and electronic counter measures (ECM). Basically, the EW community takes as its job the degradation of radar and telecommunication capabilities. The radar community takes as its job the successful application of radar in spite of what the EW community does; the goal is pursued by means of electronic counter–counter measures (ECCM) techniques [3].

ESM is that division of EW involving actions taken to search for, intercept, locate, record, and analyze radiated EM energy with the purpose of exploiting such radiations in the support of military operations. Thus, ESM provides a source of EW information required to conduct ECM, threat detection, warning, and avoidance. ECM is that division of EW involving actions taken to prevent or reduce a radar's effective use of the EM spectrum. ECCM comprises those radar actions taken to ensure effective use of the EM spectrum despite the enemy's use of EW [3].

Today the terminology has changed a bit as follows. Electronic warfare includes three major subdivisions: EA, electronic protection, and electronic warfare support (ES) [3].

8.3 Photonics in surveillance phased array radar

Before going into the details of the role—along time—of optronics in PAR it is worth recalling the several progresses that have been done along the years in developing PAR.

8.3.1 A review of beam-forming network technology and architectures

In traditional phased array beam forming, shown in Figure 8.1, the signals gathered by the array antenna are properly shifted in phase, summed in an analogue/microwave device, down-converted to the proper frequency bands, and successively transformed into digital worlds. By this well-proven technique—based on passive phased array antennas and analogue/microwave technologies—a limited number (<10 in general) of contemporary beams can be formed at radio frequency (RF) or intermediate frequency (IF) stage. The disadvantages of this architecture are related to the difficult control of the sidelobe level, high loss, absence of individual beam shape control, complex construction, and correspondingly heavy equipment.

An early digital beam-forming network, for adaptive rejection of e.m. interference and beam formation and shaping, is illustrated in Figure 8.2. In particular, the composition of each receiving channel is the cascade of (1) the RF low-noise amplifier; (2) the down-conversion from RF to IF, the bandpass filter, and the IF amplifier; (3) the base band conversion and phase detection to produce the in-phase (I) and in-quadrature (Q) channels; (4) the A/D conversion; and (5) the digital network for the formation of M beams. We should note that the number M of beams is generally different from the number of receiving antennas. In this general case,

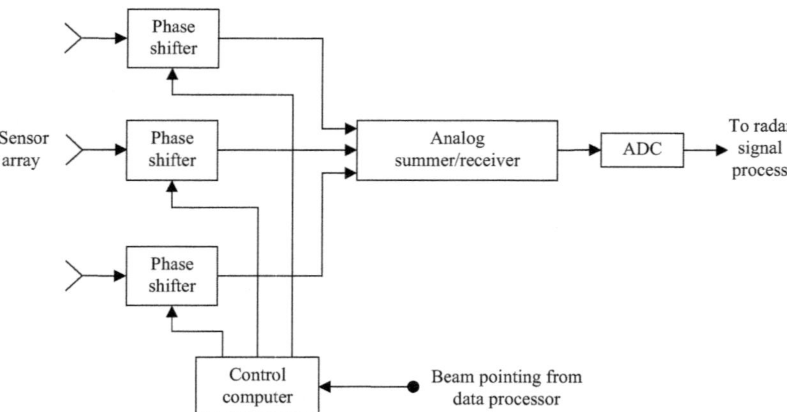

Figure 8.1 Traditional beam forming for PAR with analogue and microwave technologies [4]

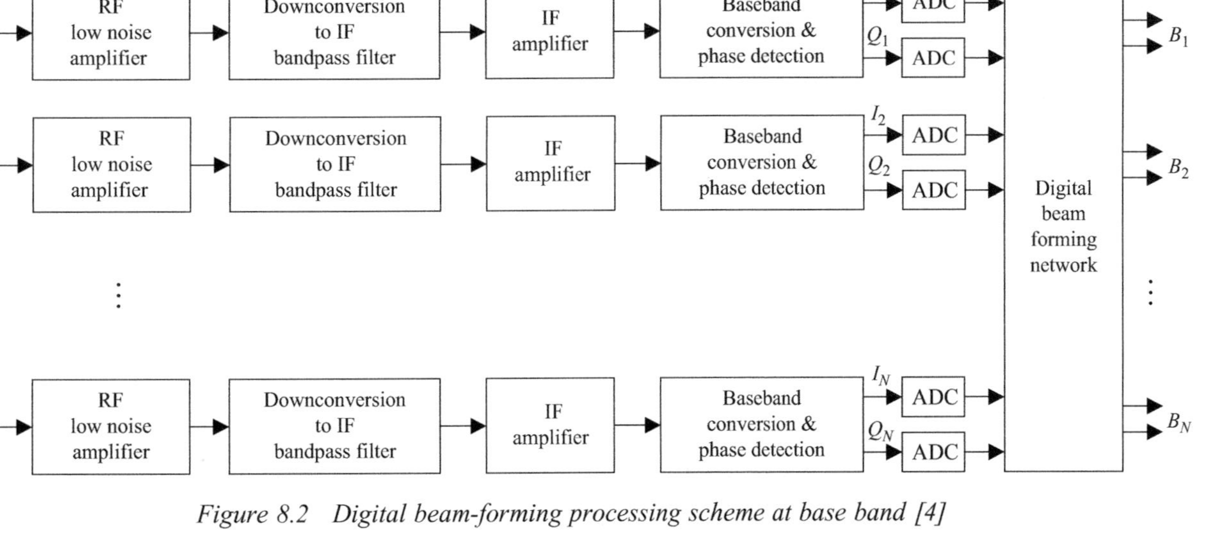

Figure 8.2 Digital beam-forming processing scheme at base band [4]

the digital beam forming is operated by a suitable (M,N) matrix $\boldsymbol{W}$ of weights, as follows:

$$\boldsymbol{B} = \boldsymbol{W}\,\boldsymbol{S} \tag{8.1}$$

where $\boldsymbol{B}$ is the M-dimensional vector of the formed beams, and $\boldsymbol{S}$ is the N-dimensional vector of samples by the N receiving channels.

Adaptivity at the subarray level has been the next architectural step for practical and realistic purposes. For an operational PAR with thousands of elements, it is not possible to adapt directly the signals from each radiating element. It is necessary to reduce the system complexity by using subarrays. A subarray is an aggregation of antenna elementary radiators; the whole antenna can be considered as an array of these super elements. Adaptive processing can be applied at the output signals of each subarray, thus reducing the system complexity. Provided that the subarrays are configured reasonably, the number of subarrays and the receiving channel errors (e.g., channel mismatching) determine the cancellation performance. Thus, the number of subarrays is a trade-off between hardware complexity, cost, and achievable performance.

In PAR it is highly desirable to have low sidelobes; this is obtained by (1) fixed weighting layer with analogue technology (i.e., at the microwave element stage) to reduce the sidelobe level everywhere, (2) fixed weights at the digital subarray level to reach a prescribed peak-to-sidelobe ratio, and (3) an adaptive weighting layer with digital technology to put nulls along the jammer direction of arrival (DoA) of high directional beams (sum, difference, cluster of high gain patterns) and low gain, possibly omnidirectional, beams (e.g., guard channel: Ω). Figure 8.3 presents a simplified scheme of a modern PAR.

Advances in RF and digital electronics are making digital arrays a reality for radar (DAR) and EW as well. As already pointed out, microwave analogue beam-forming manifolds are designed for a specific application, with increasing

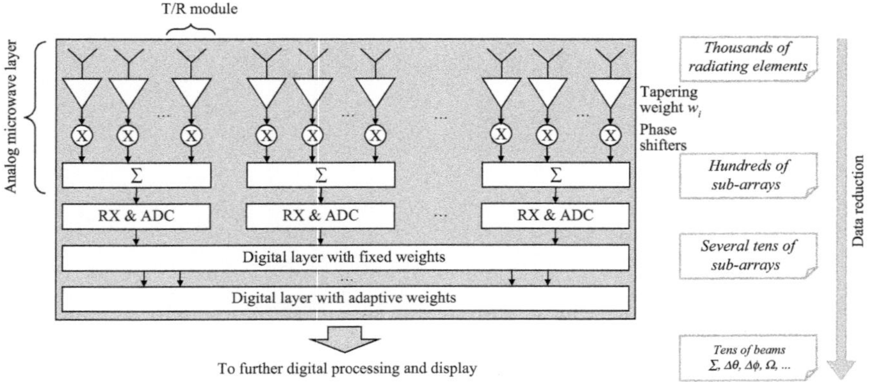

Figure 8.3 A PAR with subarrays and digital beam forming deterministic and adaptive [5]

development cycle duration, and highly complex and expensive when low antenna sidelobes are desired. Instead, digital arrays provide the opportunity for faster, lower-cost development of family of sensor system products sharing common designs and hardware, also leveraging the scalability inherent in having complete transceiver functionality at each array element [6].

For DAR applications, having direct access to the data at each array element provides flexibility and real-time reconfigurability unattainable with conventional architectures, which require that digital signal processing being after beams of a predetermined type have already been formed. Besides the flexibility, element-level digitization provides the opportunity for new and unconventional signal processing techniques [6]. Figure 8.4 gives a summary view of receive block diagrams for four main radar architectures.

The main operational advantages of DAR are as follows:

- improved adaptive pattern nulling,
- multiple simultaneous beams over the full scan volume,
- ultra-low sidelobes, and
- flexible radar power and time management.

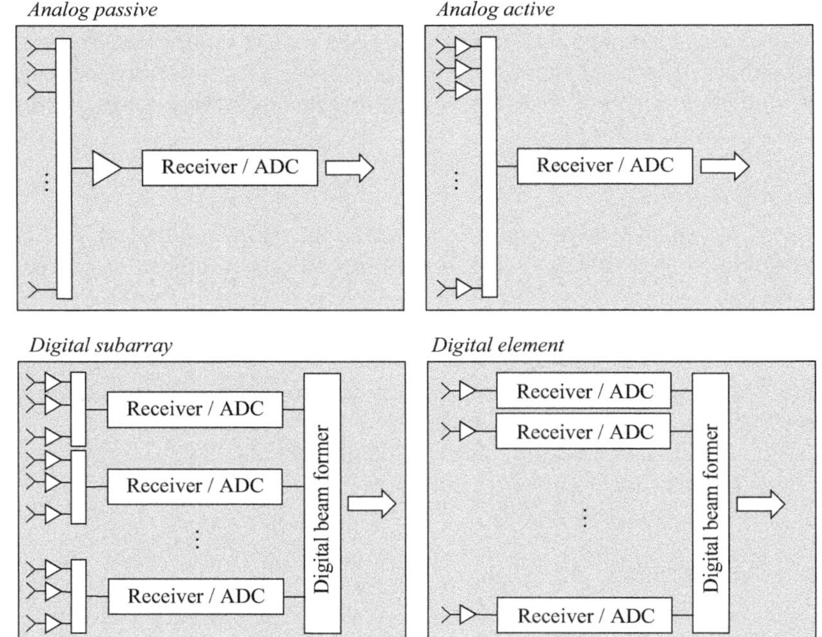

Figure 8.4 A summary view of receiver block diagrams for four main architectures: the digital element corresponds to the modern digital array radar (DAR) [6]

Moreover, it achieves enhanced motion stability due to phase noise averaging over *N* independent local oscillators. This brings to enhancement of system stability due to phase noise averaging over *N* independent local oscillators. This brings to enhancement of low RCS target detection performance in heavy clutter environment [7].

8.3.2 Past

The role of surveillance radar including PAR in the 1980s is described in [8], which highlights the state of art, research, and perspectives. Details on the signal processing for target extraction from clutter, radar networking, etc., are discussed in [9]. The special topic of ECCM was the subject of [4].

Photonics and optronics played a role in the fiber optic digital links currently used in radar local area network (LAN) and in site tools for antenna remoting or transponders (i.e., emulation of target echoes).

As a suitable reminder, we shortly recall the basic definitions of photonics and optronics considered in this chapter.

Photonics [10] is the physical science of light (photon) generation, detection, and manipulation through emission, transmission, modulation, signal processing, switching, amplification, and detection/sensing. The term photonics developed as an outgrowth of the first practical semiconductor light emitters invented in the early 1960s and optical fibers developed in the 1970s.

Optronics [11] is the study and application of electronic devices and systems that source, detect, and control light, usually considered a subfield of photonics. It is interesting to note that optoelectronics is based on the quantum mechanical effects of light on electronic materials, especially semiconductors, sometimes in the presence of electric fields.

8.3.3 Present

The role of surveillance radar, and PAR in particular, today is described in [12,13], which highlights the technology and signal processing algorithms. Details on the signal processing for target extraction from clutter, radar networking, etc., are discussed in [9]. The evolution of ECCM is reported in [14].

Fiber optic is a technology currently used in multiple analogue and digital signal (LAN, timing, and control) transport through collector in rotating antenna, in massive data transport for digital beam forming from antenna backend toward processor and in signal processing. At each antenna module, the fiber optic can feed the commands and timing and the stable local oscillator (STALO) fiber optic can transmit the radar-received data.

8.3.4 Near future

The future of PAR, not only for surveillance, is reported in the plenary talk and the special session "European Phased Array Systems and Technology" (chairmen: A. Farina, P. van Vliet) at the 2016 IEEE International Symposium on Phased Array Systems and Technology, 18–21 October 2016, Waltham, MA, USA [15]. PAR systems continue to be a rapidly evolving technology with steady advances motivated by the challenges presented to modern military and commercial applications. The symposium presented the most recent advances in phased array technology and provided a unique opportunity for members of the international community to interact with colleagues in the field of phased array systems and technology.

Recently, Dr. J. Herd (MIT-LL) released an audio interview to Andrew Zai on low-cost phased array technology for surveillance, air traffic control, and meteorology (see http://array2019.org/podcast-episode-2-interview-with-conference-chair-jeff-herd-on-low-cost-phased array).

The role of photonics/optronics has become the following. STALO can be generated in optics (optoelectronic oscillator), the integrated optics transmitter and receivers for multifunction system can be implemented in integrated optics and particularly using a mode locked laser as a coherent clock for all the system functions (multiple carrier tunable exciter, high-speed RF to digital conversion).

8.4 Photonics/optronics in SAR

8.4.1 SAR, a short reminder

SAR is a form of radar that is used to generate images of objects, such as landscapes. These images can be either two- or three-dimensional representations of the object. SAR uses the motion of the radar antenna over a targeted region to provide finer spatial resolution than is possible with conventional beam-scanning radars. SAR is typically mounted on a moving platform such as an aircraft or spacecraft, and has its origins in an advanced form of side-looking airborne radar. The distance the SAR device travels over a target in the time taken for the radar pulses to return to the antenna creates the large "synthetic" antenna aperture (the "size" of the antenna). As a rule of thumb, the larger the aperture is, the higher the image resolution will be, regardless of whether the aperture is physical (a large antenna) or "synthetic" (a moving antenna)—this allows SAR to create high-resolution images with comparatively small physical antennas [5,16,17].

8.4.2 Past

The generation of images from the signals received by SAR has an interesting technology history. Real-time SAR image formation by optical processing was performed by two fast Fourier transform in cascade. Figure 8.5 shows the optical apparatus used in 1970.

8.4.3 Present and future

The evolution of spaceborne radar payloads relies on modular designs toward System on Chip policies and ancillary message passing among periphery entities comprising manifold subsystems. Indeed payload subsystems depend on several send/receive functions aimed at programming, processing, storing, controlling, and transferring data. Moreover, from an industrial perspective it appears mandatory to foster standard procedures aimed at reducing design, development, and integration costs as well as promoting either backward compatibility or milder forms of tailoring toward legacy approaches. High-speed interfaces beyond SpaceWire and WizardLink such as Spacefibre will become keystones for modern spaceborne SAR

Figure 8.5 Tilted plane optical SAR processor. Courtesy of Dr. D. Ausherman, General Dynamics Advanced Information Systems

payload architectures in order to cope with inevitable delays for transferring data among subsystems as well as querying mass memory databases [18].

Accordingly, Spacefibre is a protocol stack designed for space applications on optical fibers over distances of up to 100 m with high throughputs on the order of tens of Gb/s as well as quality of service and fault detection isolation and recovery capabilities under the wing of the European Cooperation for Space Standardization standardization processes and related working groups [19].

Nevertheless, it is also worth noting that so far STAR-Dundee has been designing dedicated Spacefibre IP VHDL for rad-hard application specific integrated circuits (ASICs) and field programmable gate arrays (FPGAs) [20].

8.4.3.1 The role of Italy in spaceborne SAR missions

Italy is at the forefront of spaceborne SAR missions, in line with the European heritage started with the participation to the Shuttle Radar Topographic Mission (SRTM) as per the close liaison among NASA-JPL, ASI, and DLR. Indeed, the SRTM was an international research effort aimed at providing both the scientific community and governments with accurate digital elevation models on a near-global scale at that time. In particular, Italy contributed with the design and development of a X-band Synthetic Aperture Radar (SAR) which was further improved with the advent of the COnstellation of small Satellites for the Mediterranean basin Observation (COSMO-SkyMed), funded by the Italian Ministry of Research and Ministry of Defence and managed by the Italian Space Agency (ASI) for both defense and civilian uses.

The space segment of the COSMO-SkyMed system includes four medium-sized satellites (orbiting at an altitude of 619 km) equipped with SAR sensors at X-band with a global coverage of the planet. Observations of an area of interest are repeated several times a day in all-weather conditions. The imagery is exploited for defense and security in Italy and other countries, seismic hazard analysis, environmental disaster monitoring, and agricultural mapping. The first satellite was launched on June 2007. The additional three ones were lunched in the following dates: second on December 2007, third on October 2008, and fourth on November 2010 [21].

Italian scientists and engineers are renowned in SAR Interferometry for 3D localization and deformation measurement. Also modern Differential Interferometry SAR techniques to process multiple acquisitions are extensively used. Displacements time series and topography can be reliably separated in a multi-dimensional space with multiple acquisitions characterized by temporal and spatial diversity. For instance, after the terrible event of L'Aquila earthquake (2009), the aftershocks were monitored by COSMO-SkyMed [22].

The aforementioned heritage is further carried on by Cosmo Skymed Second Generation, along with several improvements in terms of radiometric and geometric performance as well as by advanced architectural designs, with operative modes improving Stripmap, Spotlight, and Scansar. Nevertheless, current efforts are also aimed at investigating swath versus resolutions trade-offs in terms of novel High Resolution Wide Swath (HRWS) modes comprising Digital Beamforming Networks (DBFN) for SCan On REceive (SCORE) in the elevation dimension jointly with either staggered Pulse Repetition Frequencies (PRF) or multichannel sampling of the azimuth spectrum. In line with these efforts, the technology readiness level (TRL) of enabling technologies related to optronics are nurtured for modular designs towards System on Chip (SoC) policies comprising manifold aspects such as multifrequency optical oscillators, optical gyros for 3-axis stabilized Attitude and Orbital Control Systems (AOCS), or Spacefibre high speed interfaces with high throughputs on the order of tens of Gbps as well as quality of service (QoS) and Fault Detection, Isolation and Recovery (FDIR) capabilities.

8.5 The role of photonics/optronics in adaptive digital beam forming (ADBF) for radar

The role of photonics/optronics in ADBF for multichannel radar has been analyzed in a study case in the early 2000s, and the achieved results were published in [23]. Here a brief summary is reported in relation to the exploitation of an optical processor. The ADBF for a multichannel radar is relevant both for surveillance radar and for SAR. Multichannel, especially Space-Time Adaptive Processing (STAP), technique was originally conceived to suppress clutter (and directional e.m. interference) received by a radar on board a moving platform, for example, an aircraft. This was so because of the clutter ridge in the Doppler frequency and angle (spatial frequency) plane; in fact, the clutter echoes are coupled in angle and Doppler frequency. Thus, a two-dimensional adaptive filter is required to cancel the disturbance [24]. When the platform is stationary, this coupling does not exist anymore. However, a STAP-like technique (i.e., joint adaptive cancellation in both frequency and angle domains) is still valuable when clutter and directional e.m. interference are present and their statistical features are not known a priori. In fact, if we use the cascade of an adaptive moving target indication (MTI) operation mode (to suppress the clutter) and an adaptive array (to suppress the directional e.m. interference) [25], it happens that both adaptive filters will not correctly work in the presence of strong interference. For instance, the adaptive MTI will not correctly estimate the clutter covariance matrix in the presence of a strong directional e.m. interference, thus it will not generate the proper filter weights to cancel the clutter. This does not allow to have clutter-free range cells to be used to estimate the directional e.m. interference covariance matrix for the disturbance cancellation [26]. In summary, the adaptive spatial array will not correctly estimate the directional e.m. interference covariance matrix in the presence of strong clutter. Thus, the only way to have a successful adaptive processor to reject both clutter and directional e.m. interference is the STAP-like, that is, the joint cancellation in the Doppler frequency angle domain. The paper [23] presents some experimental results obtained in the application of STAP-like algorithms to the live data recorded using a test-bed microwave multichannel radar during a suitable experimental campaign. The STAP-like algorithm is implemented by emulating a systolic array of processing elements, which performs a QR decomposition on the clutter, e.m. interference, and noise live data. The constraints for the target Doppler frequency and the target direction of arrival we are looking at are imposed resorting to the minimum variance distortionless response approach. Details on the algorithms are found in [26–28].

The performance advantages of the STAP-like algorithm shall be traded-off with the computational cost to allow the STAP application in current radar systems. The paper [23] presents a possible strategy to use STAP with a limited increase of the computational burden. Two approaches to achieve the computation load required by the STAP are presented. The first refers to the mapping of the QR decomposition on an FPGA-based device. The mapping has been tested with simulated data relative to a linear array with five antenna dipoles; this is for the e.m. directional interference cancellation along the angle domain. The second approach refers to the mapping of the STAP algorithms on a high-speed optical processor. The basic operation of this processor is vector–matrix multiplication in a single step (256-element vector by a 256×256 matrix) at a rate of 125 MHz. The optical processor, designed by Lenslet Company, was capable (years 2000s) of performing 8,000 Giga multiply and accumulate operations per second and therefore is suitable for computational-intensive applications such as STAP. The approach has been tested with simulated data for a linear array with 10 antenna dipoles each with 10 pulse repetition time intervals. The interference to signal power ratio tested in this case was 80 dB.

This section contains the description of the mapping of an ADBF algorithm on an optical-based computing device and comparison with FPGA performance. More details are in [23]. Today Lenslet Company is still operating on the field as reported in [29]. Quoting from the website:

> The EnLight electro-optic processor operates at the unprecedented speed of 8 tera (8,000 billion) calculation operations per second – one thousand times faster than any known digital signal processor. A small Israel company called Lenslet has developed a revolutionary electro-optic processor which operates at the unprecedented speed of 8 Tera (8,000 billion) calculation operations per second – one thousand times faster than any known Digital Signal Processor.

8.5.1 Mapping of an ADBF algorithm on optical computer

8.5.1.1 Optical processor description

Optical processing is inherently capable of high parallelism, which can be translated to very fast computational power. Lenslet developed the world's first commercial optical processor. Lenslet processor, the EnLight256 is a small factor signal processing chip ($5 \times 5\text{cm}^2$) with an optical core that performs 8 tera multiply-accumulate operation (MAC)/second; that is, 8,000 billion calculation operations per second. This is 1,000 times faster than the fastest digital signal processor (DSP) available today (i.e., at the date of publication of [23]). The optical core is a vector–matrix multiplier (VMM), with a matrix size of 256×256. The system clock is 125 MHz. At each clock, 256^2 MACs are carried out, resulting in the 8 TeraMACs per second performance figure. The scaling and parallelism properties of an optical processor are behind the matrix size choice—the larger is the scale the faster the computation, with relatively small scaling penalty, comparing to electronics.

In addition to the optical VMM, the EnLight 256 optical processor includes a vectoric processing unit (VPU), which deals with vector–vector operations. The combination of the VMM and VPU units provides a powerful computing platform for complex algorithms.

8.5.1.2 Description of mapped algorithm

The challenge of using the optical processor for resolving STAP-heavy computational tasks was to find innovative ways to cope with stringent requirements for dynamic range taking into account the specific characteristics of the OP and supplied data. The OP is characterized by extremely high performance for the matrix–vector product operation, limited input/output accuracy and preference to a slower matrix update rate than the vector speed.

The supplied data was considered as a set of sample vectors, for example, $(0.5 \div 1) \cdot 10^3$ vectors. The length of single data sample vector was about 10, for example, 10 antennas $\times$ 10 time taps. The covariance matrix was highly singular, with the condition number $\sim 10^8$.

In order to handle such extreme characteristics, an algorithm was developed by Lenslet, which worked in the data domain. It was a particular combination of Jacobi–Davidson algorithm and method based on Krylov subspace iterations.

The algorithm consisted of the following steps:

1. Calculation of the Krylov subspace basis for augmented matrix. This was performed on VMM and VPU. This was the only step that demanded full matrix–vector product calculation. In parallel to the process of calculating the subspace basis, the Jacobi-type orthogonalization matrix (tri-diagonal symmetric real) was calculated as well.
2. Calculation of the Ritz approximation for the eigenvectors and eigenvalues of the covariance matrix. This was performed on DSP and VPU. First, the eigenproblem was solved for the tri- diagonal Jacobi matrix obtained in the previous step. The eigenvectors were obtained as a linear combination of the Krylov subspace basis. The results served to estimate the corresponding elements of the interference covariance matrix.
3. Calculation of the filter coefficients. This was performed on VPU according to Hung–Turner projection algorithm [30].
4. Calculation of the interference residual for each steering vector (looking direction).

8.5.1.3 Algorithm performance

The time performance was proportional to the dimension of the Krylov subspace (step 1 of the algorithm). To calculate the single basis vector, the following operations had to be performed sequentially:

1. Multiplication of the augmented matrix over a vector. The practical implementation of this operation required calculation of matrix–vector products for two matrices: one was the part of the row data matrix taken for covariance matrix estimation, and the second one was the transposed matrix.
2. Dot product of two vectors.
3. Product of a vector and a scalar.
4. Subtraction of two vectors.
5. Normalization of a vector.

Denoting the length of the data vector by N ($N = n \cdot m$) and by K the number of sample vectors to be processed, the number of operations necessary to calculate the single basis vector is ~$2NK$.

8.5.1.4 Implementation and time performance of a notional system

The algorithm performance and the influence of the constraints of the OP limited accuracy operations were tested on the EnLight256 simulator written on MATLAB© for the supplied data for the study case.

The data set included 700 vectors, each one of length 100 (10 antenna × 10 taps). About 20 eigenvectors of the covariance matrix were of the order $10^7 - 10^8$ while the rest were of the order 1–10. The results are presented in Figure 8.6. The green line ("augmented Krylov") corresponds to the interference residual provided by EnLight

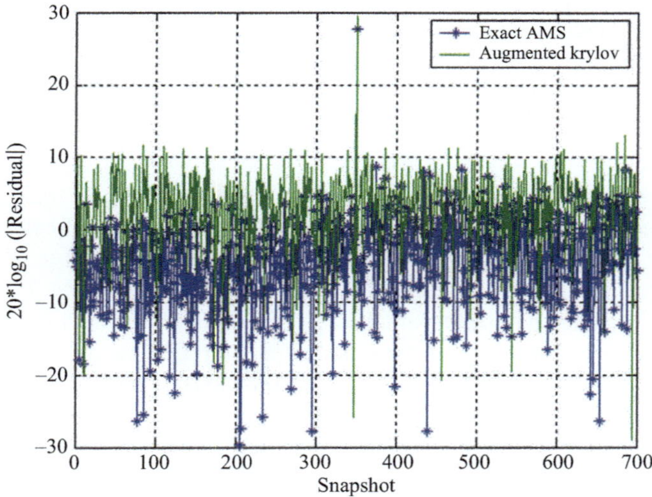

Figure 8.6 10 antennas and 10 taps STAP results on the optical processor compared to MATLAB©

simulator. The blue line with asterisks shows the exact results. The results were obtained with the Lenslet system, which had 10 bits accuracy for the input vector, 10 bits for the matrix inputs, and 10 bits for the output vector, that is, the notional 10–10–10 system.

The time required for the basic calculation of the Krylov subspace was proportional to its dimension. It required two clock cycles to perform a single basis vector calculation. In the above test case example, the dimension of the Krylov subspace was 60, therefore 120 clocks were required to calculate it.

In order to achieve the 10 bits accuracy, identical calculations were repeated 32 times. Therefore, in total, the calculation for 60 vectors of the subspace took:

$$60 \times 2 \times 32 = 3,840 \text{ VMM clock cycles @ } 125 \text{ MHz} = 31 \ \mu\text{s}.$$

The next step in the algorithm was to complete the calculation of the Ritz approximation of the eigenvectors and the eigenvalues of the covariance matrix and then to calculate the filter coefficients. To accommodate these calculations, and with an assumption of limited pipeline utilization of the VPU due to the recursive nature of this algorithm, Lenslet estimated that 20 μs was needed for the additional VPU overhead. Therefore, the estimate for the complete calculation of weights for 100 different steering vectors was an impressive 50 μs. Since this method works on the data domain, there is no time spent on calculating the covariance matrix.

8.5.1.5 Optical versus FPGA performance

The case studies, which have been examined for the FPGA and optical processor, are different in their size and scale due to the significant performance gap between the two devices. In order to provide a rough estimation of the performance ratio between the two devices, when used for radar STAP systems, Lenslet has estimated the type of solution it can provide within the same 0.4 μs (2.5 MHz) that has been tested in the FPGA. Table 8.1 shows the result of this comparison.

Lenslet's estimations provided in Table 8.1 have used a dual EnLight (ODSP) configuration. Assuming N^2 computational complexity, we obtain that for STAP implementations a pair of Lenslet's EnLight optical processors is equivalent to 160 FPGAs. We conclude that the implementations of a STAP in future radar systems may become a key technological differentiator. Optical processors can serve as an enabling building block for larger antenna arrays with more time taps in reasonable physical dimensions and overall system costs.

Table 8.1 EnLight optical processor versus FPGA

	FPGA	EnLight ODSP
Antennas	5	8
Taps	1	8
N	5	64

8.6 The role of photonics/optronics in ESM

Optronics is widely exploited in several weapon systems (e.g., classical missile, man pad, air to air with infrared search) and in counter measures systems like the Directional Infrared Counter Measures (DIRCM). The latter is a system to protect aircraft from infrared homing ("heat seeking") man-portable missiles. It is a lightweight, compact system designed to provide mission-vulnerable aircraft with increased protection from common battlefield threats. It is more advanced than conventional infrared countermeasures. The term DIRCM is used as a generic term to describe any infrared countermeasure system that tracks and directs energy toward the threat [31].

It can be integrated with whatever EW at RF device. Thus, the optronic technology is an important part of today EW.

8.7 Coexistence of radar and EW: the role of optronics

The receiver operating characteristics (ROC) (detection, false alarm probabilities, and signal to noise power ratio) define, among other parameters, the radar performance [32]. ESM performance is defined by the probability of intercept, the time of intercept, and the probability to recognize a radiating source [33]. The two systems when suitably collocated on a platform should be properly synchronized in time, frequency, and look angle to avoid improper function of each other. Suitable spatial separation and antenna setting can help the e.m. uncoupling of the two systems. A system manager should harmonize the operational functions of each system and the performance of the integrated whole [34].

Optronics could play a role for collection and distribution of data as well as provide powerful signal processing computing devices.

8.8 Quantum sensing and quantum radar (QR): Sci-Fi or a potential reality?

Quantum radar is a remote sensing method based on quantum entanglement[1] [36]. The most convincing model has been proposed by an international team of researchers [37].

[1]*Quantum entanglement* is a physical phenomenon that describes pairs or groups of particles where the quantum state of each particle cannot be described independently of the others. The whole system including the particles must be described by a quantum state. The entanglement affects the measurements of physical properties (e.g., position, momentum, spin, or polarization) of entangled particles. For example, let's suppose that a pair of particles are generated with a total spin equal to zero. If one particle is measured to have clockwise spin on a certain axis, the spin of the other particle on the same axis will be measured to be counterclockwise. Therefore, any measurement of a property of an entangled particle acts on that particle (e.g., by collapsing a number of superposed states), and on the entangled system as a whole. One particle of an entangled pair "knows" what measurement has been performed on the other, and with what outcome, even if there is not any known means of communication between the particles, which at the time of measurement may be separated by arbitrarily large distances [35].

This team designed a model of quantum radar for remote sensing of a low-reflectivity target that is embedded within a bright microwave background, with detection performance well beyond the capability of a classical microwave radar. By using a suitable wavelength converter, this scheme generates excellent quantum correlations (quantum entanglement) between a microwave signal beam, sent to probe the target region, and an optical idler beam, retained for detection. The microwave return collected from the target region is subsequently converted into an optical beam and then measured jointly with the idler beam. Such a technique extends the powerful protocol of quantum illumination (QI) [38] to its more natural spectral domain, namely microwave wavelengths.

A prototype quantum radar could be realized with current technology and is suited to various potential applications, from standoff sensing of stealth objects to environmental scanning of electrical circuits. Thanks to its quantum-enhanced sensitivity, this device could also lead to low-flux noninvasive techniques for protein spectroscopy and biomedical imaging.

Alternative methods were also considered by defense contractor whose aim was to create a radar system providing a better resolution and higher detail than classical radar could provide [39].

According to Chinese state media, the first quantum radar was developed and tested in real-world environment in August 2016 [40]. This story originally was sourced from the Chinese government's very own paper the *Global Times* [36].

8.8.1 Basic principle of operation of QR

As briefly described in [39] the basic idea is to have two photons one on the aircraft to detect and the other one on the radar receiver as illustrated in Figure 8.7.

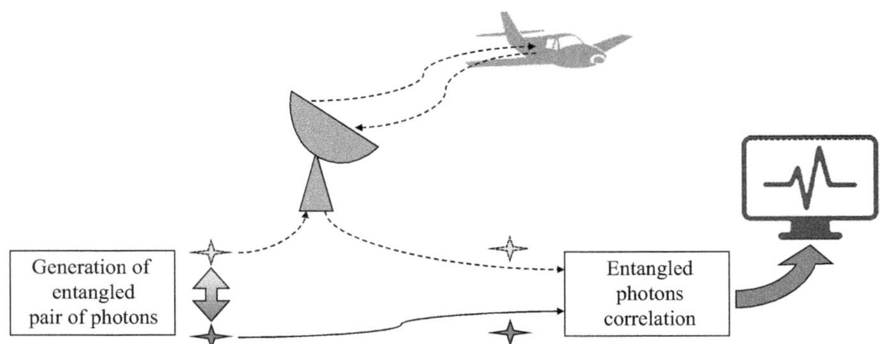

Figure 8.7 The working principle of quantum radar [39]: a pair of entangled photons is generated, one of the photons is sent toward the target, and once it is reflected back to the radar, it is correlated with the other photon, increasing the sensitivity of the system

8.8.2 Basic principle of operation of QI

S. Lloyd and the international working group have set up the technique of quantum entanglement between photons and microwaves [37,38], called QI, which allows the QR to be a potential real system.

The QI target-detection system operates in the microwave regime, just as any radar, generating entangled microwave and optical signals (see Figure 8.8). The main building block is the electro-optomechanical (EOM) converter (Figure 8.8(a)), where a mechanical resonator is shared among a microwave cavity and an optical cavity. This way, the EOM converter can sustain simultaneously both a microwave oscillation and an optical oscillation, with these two emitted signals being entangled.

The QI scheme is depicted in Figure 8.8(b). The transmitting EOM converter generates the entangled microwave and optical signals. The microwave signal is launched toward the target, while the optical (idler) signal is kept for the detection, as will be described in the following. The microwave echo signal reflected by the target is collected at the receiving EOM converter, where it drives the microwave cavity to oscillate, and thus the optical cavity of the receiving EOM converter oscillates too. Finally, the optical signals from the transmitting and receiving EOM converters are correlated in an optical detector.

The authors claim that their system dramatically outperforms a conventional (coherent-state) microwave radar of the same transmitted energy, achieving an orders-of-magnitude lower detection-error probability. Moreover, the system can be realized with state-of-the-art technology and is suited to such potential applications as standoff sensing of low-reflectivity objects, and environmental scanning of electrical circuits. Thanks to its enhanced sensitivity, the system could also lead to low-flux noninvasive techniques for protein spectroscopy and biomedical imaging [37].

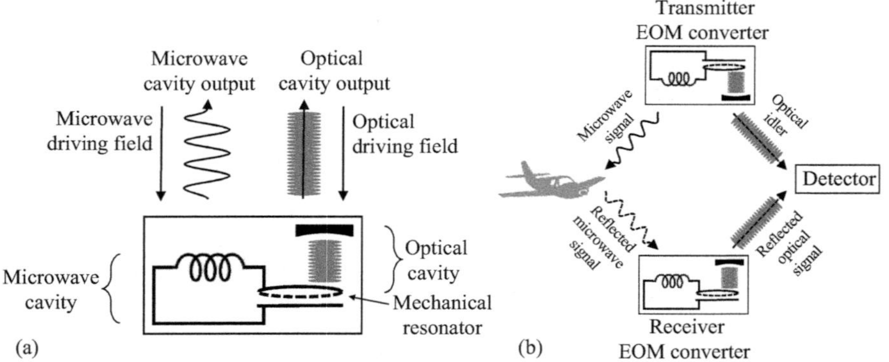

Figure 8.8 (a) Scheme of the electro-optomechanical (EOM) converter; (b) the two OEM converters in the transmission and reception phases (adapted from [37])

The key technology of the QR is the tool EOM converter, which produces an entanglement between an optical signal and a microwave signal (see Figure 8.8). One could argue that the two OEM in the transmission section and reception section should be twins; otherwise, their difference will heavily affect the radar performance.

Moreover, as stated in [37], the main goal of the QI was to boost the detection of a low-reflectivity target immersed in a bright thermal background. We note that the targets of interest are in general at a high altitude where the temperature may be quite low. Thus, this type of QR could be in principle applied to detect target at very short ranges. The paper itself mentions the sensing of electric circuits to scan, spectroscopy, and biomedical imaging.

Another limitation seems related to the optical store of the idler mode to be preserved during signal round trip time. For long round-trip time, it has been evaluated a loss of 3 dB, implying no advantage with respect to the classical radar.

It is to be noted that the target detection occurs by comparing the number of photons collected by the system in the two alternative hypotheses H1 versus H0. This processing mode is understandable even though it is formally different from the way the classical radar operates.

A key question refers to where is the transmitter located: Is it the mechanical resonator of the microwave cavity of Figure 8.8? Is it possible that such a small power is enough to transmit?

Additional comments to [37] follow. The detection performance figure of radar—the classical ROC—shown as figure 3 in [37] is not the typical one of classical radar; on the x-axis we do not put the time-bandwidth product, rather the signal to noise power ratio! An *equation of quantum radar* seems missed. It is worthy to have and compare it with that of classical radars.

It is argued that the QR looks at a range-angle-Doppler cell once at a time. Finally, the study reported in [37] is done just in the presence of noise. How the clutter, multipath, and other disturbing phenomena affect the optical-microwave entanglement?

In summary, it seems that the technology of QR is still in its infancy; nonetheless, it is worth mentioning and keeping it under track. A potential proof of concept of QR with practical limitations and potential applications is certainly interesting to have. This statement has been confirmed recently by a plenary talk of S. Pirandola at EuRAD2018, Madrid, 23–25 September: "Quantum radar: from quantum illumination to a working prototype," where a potential prototype can provide few kilometer detection range on a target.

8.9 Summary and way ahead

To sum up, it has been shown that photonics plays a key role in the RADAR and EW systems. When we speak about technologies, especially the new ones, it seems necessary to quote/estimate their Technology Readiness Level (TRL) [41]. TRL is a method of estimating technology maturity of critical technology elements of a program during the acquisition process. This is an open point that deserves further investigation.

References

[1] http://www.mitre.org/publications/systems-engineering-guide/se-lifecycle-building-blocks/concept-development/operational-requirements. Accessed date: 2018-11-14.

[2] E. J. Baranoski, "Through-wall imaging: historical perspective and future directions", *Journal of the Franklin Institute*, vol. 345, no. 6, pp. 556–569, 2008.

[3] "Electronic Warfare", Wikipedia, https://en.wikipedia.org/wiki/Electronic_warfare. Accessed date: 2018-11-14.

[4] A. Farina, *Antenna Based Signal Processing Techniques for Radar Systems*. Artech House, Inc., Norwood (MA), USA, 1992.

[5] Synthetic aperture array, Wikipedia, https://en.wikipedia.org/wiki/Synthetic_aperture_radar. Accessed date: 2018-11-14.

[6] S. H. Talisa, K. W. O'Haver, T. M. Comberiate, M. D. Sharp, and O. F. Somerlock, "Benefits of digital phased array radars", *Proceedings of the IEEE*, vol. 104, no. 3, pp. 530–543, 2016.

[7] J. S. Herd and M. D. Conway, "The evolution to modern phased array architectures", *2015 Proceedings of the IEEE*, vol. 104, no. 3, pp. 519–529, 2016.

[8] A. Farina and G. Galati, "Surveillance radars: state of art, research and perspectives". Alta Frequenza, no. 4, vol. LIV, 1985. pp. 261–277, invited paper. (Reprint on *Radar Applications*, Editor M. I. Skolnik, IEEE Press, 1988, paper no. 3.3).

[9] A. Farina (Ed.), *Optimized Radar Processors*. On behalf of IEE, Peter Peregrinus Ltd., London, October 1987.

[10] "Photonics", Wikipedia. https://en.wikipedia.org/wiki/Photonics. Accessed date: 2018-11-14.

[11] "Optronics", Wikipedia. https://en.wikipedia.org/wiki/Optoelectronics. Accessed date: 2018-11-14.

[12] A. De Maio, A. Farina, L. Timmoneri, and M. Wicks, "Ground-based early warning radar (GBEWR): technology and signal processing algorithms", chapter 8, *Principles of Modern Radar: Radar Applications*, vol. 3, W. L. Melvin and J. A, Scheer (Eds.), Raleigh, NC, USA: Scitech Publishing, an imprint of the IET, 2014, pp. 323–381.

[13] A. Farina, P. Holbourn, T. Kinghorn, and L. Timmoneri, "AESA radar – pan-domain multi-function capabilities for future systems", Invited to the plenary session at 2013 IEEE International Symposium on Phased Array Systems & Technology Boston, 15–18 October 2013.

[14] A. Farina, "Electronic counter-countermeasures", chapter 24, *Radar Handbook*, 3rd edition, M. I. Skolnik (Ed.), Mc-Graw Hill, Inc, USA, January 2008.

[15] https://ieeexplore.ieee.org/xpl/mostRecentIssue.jsp?punumber=7813534. Accessed date: 2018-11-14.

[16] D. A. Ausherman, A. Kozma, J. L. Walker, H. M. Jones, and E. C. Poggio, "Developments in radar imaging," *IEEE Transactions on Aerospace and Electronic Systems*, vol. AES-20, no. 4, pp 363–400, 1984.

[17] A. Farina, "Radar imaging: an industrial point of view. From the beginning to its applications to large systems", GTTI 2011 meeting, Messina-Taormina, 21 June 2011.

[18] http://2016.spacewire-conference.org/programme/. Accessed date: 2018-11-14.

[19] http://spacewire.esa.int. Accessed date: 2018-11-14.

[20] http://www.star-dundee.com. Accessed date: 2018-11-14.

[21] https://en.wikipedia.org/wiki/COSMO-SkyMed. Accessed date: 2018-11-14.

[22] D. Reale, D. O. Nitti, D. Peduto, R. Nutricato, F. Bovenga, and G. Fornaro, "Postseismic deformation monitoring with the COSMO/SkyMed Constellation", *IEEE Geoscience and Remote Sensing Letters*, vol. 8, no. 4, pp. 696–700, 2011.

[23] A. Farina, S. Stefanini, L. Timmoneri, *et al.* "Multichannel radar: advanced implementation technology and experimental results", *International Radar Symposium - IRS* 2005, 6–8 September 2005, Berlin, Germany.

[24] R. Klemm, *Principles of Space-Time Adaptive Processing*, The Institution of Electrical Engineers, London, UK, 2002.

[25] A. Farina, G. Golino, and L. Timmoneri, "Comparison between LS and TLS in adaptive processing for radar systems", Proceedings of IEE Radar, *Sonar and Navigation*, no. 1, pp. 2–6, 2003.

[26] L. Timmoneri, I. K. Proudler, A. Farina, and J. C. McWhirter, "QRD-based MVDR algorithm for multipulse antenna array processing", *Proceedings of IEE Radar, Sonar and Navigation*, no. 2, pp. 93–102, 1994.

[27] A. Farina and L. Timmoneri, "Real-time STAP techniques", *IEE ECEJ Special Issue on Space-Time Adaptive Processing*, vol. 11, no. 1, pp. 13–22, 1999.

[28] P. Bollini, L. Chisci, A. Farina, M. Giannelli, L. Timmoneri, and G. Zappa, "QR versus IQR algorithms for adaptive signal processing: performance evaluation for radar applications", *Proceedings of IEE Radar, Sonar and Navigation*, vol. 143, no. 5, pp. 328–340, 1996.

[29] https://www.israel21c.org/new-israeli-electro-optic-processor-is-as-fast-as-a-super-computer/. Accessed date: 2018-11-14.

[30] J.R. Guerci, *Space-Time Adaptive Processing for Radar*, Artech House, USA, p. 134, 2003.

[31] https://en.wikipedia.org/wiki/Directional_Infrared_Counter_Measures. Accessed date: 2018-11-14.

[32] M. I. Skolnik, *Introduction to Radar Systems (Irwin Electronics & Computer Engineering)*, 3rd edn., 2015.

[33] R. C. Wiley, *ELINT: The Interception and Analysis of Radar Signals*, Artech House, USA, 2006.

[34] S. Celentano, A. Farina, G. Foglia, and L. Timmoneri, "Co-existence of AESA (active electronic scanned array) radar and electronic warfare (EW)

systems on board of a military ship", SeaFuture2018, www.seafuture2018.it, 19–23 June, La Spezia.

[35] https://en.wikipedia.org/wiki/Quantum_entanglement. Accessed date: 2018-11-14.

[36] https://en.wikipedia.org/wiki/Quantum_radar. Accessed date: 2018-11-14.

[37] S. Barzanjeh, S. Guha, C. Weedbrook, D. Vitali, J. H. Shapiro, and S. Pirandola, "Microwave quantum illumination", *Physical Review Letters* vol. 114, no. 080503, pp. 1–5, 2015. DOI: 10.1103/PhysRevLett.114.080503.

[38] S. Lloyd, "Enhanced sensitivity of photo detection via quantum illumination", *Science* vol. 321, pp. 1463–1465, 2008.

[39] M. Lanzagorta, *Quantum Radar*, San Rafael, CA, USA: Morgan & Claypool, 2011.

[40] http://phys.org/news/2015-02-big-future-quantum-radar.html. Accessed date: 2018-11-14.

[41] https://en.wikipedia.org/wiki/Technology_readiness_level. Accessed date: 2018-11-14.

Further reading

A. Bergeron, L. Marchese, M. Doucet, *et al.*, "Rugged SAR optronic SAR processing through wavefront compensation and its digital analogy", *Synthetic Aperture Radar, 2012. EUSAR. 9th European Conference on.* VDE, Nürnberg, Germany, pp. 746–748.

L. Chisci and A. Farina, "Survey on estimation", *Encyclopedia of Systems and Control*, edited by T. Samad and J. Baillieul, Springer-Verlag, London, pp. 24, 2015,. DOI 10.1007/978-1-4471-5102-9_60-2.

C. W. Helstrom, "Quantum detection and estimation theory", *Journal of Statistical Physics*, vol. 1, no. 2, pp. 231–252, 1969.

C. W. Helstrom, *Quantum Detection and Estimation Theory*, Academic, New York, 1976.

M. Lanzagorta and J. Uhlmann, "Quantum computer science", *Synthesis Lectures on Quantum Computing* #2, Morgan & Claypool Publishers, 2009.

L. Marchese, M. Doucet, B. Harnisch, *et al.*, "Full scene SAR processing in seconds using a reconfigurable optronic processor", *IEEE Radar Conference*, 2010, Arlington, VA, USA, pp. 1362–1364.

S. Melo, S. Maresca, S. Pinna, *et al.*, "High precision displacement measurements in presence of multiple scatterers using a photonics-based dual-band radar", *IET International Radar Conference*, Glasgow (UK), 23–26 October 2017.

L. Pierno, M. Dispenza, G. Tonelli, A. Bogoni, P. Ghelfi, and L. Poti, "A photonic ADC for radar and EW applications based on modelocked laser", *Microwave Photonics*, 2008, Gold Coast, Australia.

L. Pierno, A. M. Fiorello, A. Secchi, and M. Dispenza, "Fibre optics in radar systems: advantages and achievements", *Polaris Innovation Journal*, vol. 22, pp. 64–71.

L. Pierno, A. M. Fiorello, A. Secchi, and M. Dispenza, "Fibre optics in radar systems: advantages and achievements", *2015 IEEE Radar Conference*, Arlington, Virginia, USA.

Selex ES Technical Review, "Photonics and more: a word of light around us", *Polaris Innovation Journal*, vol. 21, 2015.

"The Quantum Age. Technological opportunities". Government Office for Science, UK, pp. 1–64. https://www.gov.uk/government/uploads/system/uploads/attachment_data/file/564946/gs-16-18-quantum-technologies-report.pdf. Accessed date: 2018-11-14. © Crown copyright 2016.

V. Tocca, D. Vigilante, L. Timmoneri, and A. Farina, "Adaptive beamforming algorithms performance evaluation for active array radars", *IEEE Radar Conference* 2018, 23–27 April, Oklahoma City, OK, USA.

S. Tonda-Goldstein, D. Dolfi, and A. Monsterleet, "Optical signal processing in radar systems", *Microwave Theory and Techniques, IEEE Transactions on*, vol. 54, no. 2, pp. 847–853, 2006.

M. Weber and D. Zrnic, "Meteorological phased array radar: opportunities, challenges and outlook", *2018 IEEE Radar Conference*, 23–27 April 2018, Oklahoma City, OK, USA.

Chapter 9
Conclusions

Antonella Bogoni[1,2], *Paolo Ghelfi*[2], *and Francesco Laghezza*[3]

As closing remarks, we think it is worth summarizing here the main points highlighted throughout this book.

In the last few decades, photonics has demonstrated allowing several functional capabilities in the microwave world: the wideband generation, detection, and distribution of microwave signals, the agile filtering of radio frequency (RF) signals, and the fast and precise control of signal phase or delay for the beamforming functionality in phased array antennas. The main features that photonics can bring to the microwave field are the frequency flexibility (allowing the operation from few gigahertz up to several tens of gigahertz with the very same devices) and the precision (in terms of phase stability, also in conjunction with the fiber distribution). These can be exploited in microwave systems to reach new performance and new functionalities.

Meanwhile, the applicative fields of radar and electronic warfare systems are facing serious problems due to the most recent performance requirements.

The radar systems shall integrate multiple capabilities, implementing the concept of cognitive multifunctional radar to reduce the overall cost of the surveillance apparatus. The availability of multiband and frequency-flexible transceivers would also open up new possibilities associated with the concept of software-defined radar.

Moreover, coherent radar networks have become a necessity for gathering more accurate and comprehensive information exploiting the angular diversity and data fusion. In this scenario, a perfect synchronization among the sensor nodes is required to implement a high-performance centralized approach.

On the other hand, electronic warfare (EW) systems need to expand their range of managed frequency up to several tens of gigahertz, reducing at the same time the size, weight, and consumption of the apparatus to fit the most demanding platforms, as the unmanned vehicles.

[1]Institute of Technologies for Communication, Information and Perception (TeCIP), Sant'Anna School of Advanced Studies, Italy

[2]National Laboratory of Photonic Networks and Technology (PNTLab), Consorzio Nazionale Interuniversitario per le Telecomunicazioni (CNIT), Italy

[3]NXP Semiconductors, The Netherlands

From the above scenario, it is self-evident that photonics can meet all these new requirements, becoming a real technological enabler. Indeed, we have reported in particular on recent prototypes of radars, radar networks, and EW receivers that exploit photonics to develop peculiar innovative features: respectively, simultaneous multiband operation, coherent multistatic and largely distributed surveillance, and ultra-broadband analysis. These features would be hard—if not impossible—to implement with standard RF technologies, while they are readily available if photonics is used instead. Moreover, it is worth recalling that these prototypes have already shown overall performance aligned with state-of-the-art commercial systems, although they are just at a demonstrator stage and are built with discrete optical components, so further significant improvements (from the viewpoints of general performance, reliability, dimensions, and costs) can be reasonably expected.

The rapid advancements of the photonic integration technologies is a pivotal point for the industrial development of microwave photonics. In fact, in several applications (in particular, in the military field) the reduction of size and weight of a system can be the main driver for inserting photonics in a purely electronic world. Moreover, as already pointed out, realizing a microwave photonics system by means of integrated technologies does not bring only a reduction of size and weight (and sometimes of power consumption and cost as well), but it also allows an improvement in performance and reliability.

Lot of efforts are being spent worldwide for improving the performance of the integration processes and of the integrated devices, and today several established technological platforms are available. Among them, silicon photonics and indium phosphide are the most relevant, but several others are there too, as the low-loss silicon nitride, or hybrid approaches that merge the benefits of different technologies in a common platform (e.g., silicon and silicon nitride). Besides being technically well established (i.e., reliable), these technologies can be accessed at affordable costs by means of multiproject runs, favoring the investigation of the photonics benefits by the microwave industry.

The packaging technologies are also of furthermore importance, in particular for complex photonic systems requiring different components that cannot be integrated in the same platform. In these cases, "co-packaging" techniques fitting different integrated devices in a single package are fundamental to provide a single, easy-to-use integrated photonics-based RF system.

Currently, the advancements on microwave photonics systems are being driven by the academic research. In order to bring these breakthrough solutions to the market, it is necessary to exploit integrated solutions. Nevertheless, the time and costs for designing, realizing, and fully developing photonic-integrated systems require a direct industrial interest and investment.

By now, industries are observing with evident interest the development of the photonic technologies in the microwave field. Although they still are not fully taking the lead, the increasing amount of demonstrations with very competitive results are pushing industries to jump in, and to accelerate the development process.

With these premises, we wish to witness soon the birth of a dynamic industrial microwave photonics sector, developing RF systems (in particular, but not limited to, radars and EW systems) with completely new technical features enabled by the flexibility of photonics, and implementing breakthrough applications in several ambits of our life.

Index